HOW TO SURVIVE THE *TITANIC*

ALSO BY FRANCES WILSON

Literary Seductions: Compulsive Writers and Diverted Readers

The Courtesan's Revenge

The Ballad of Dorothy Wordsworth: A Life

HOW TO SURVIVE THE *TITANIC*

The Sinking of J. Bruce Ismay

Frances Wilson

HARPER

An Imprint of HarperCollins*Publishers*

www.harpercollins.com

HarperCollins books may be purchased for educational, business, or sales promotional use. For information, please write: Special Markets Department, HarperCollins Publishers, 10 East 53rd Street, New York, NY 10022.

First published in Great Britain in 2011 by Bloomsbury.

FIRST U.S. EDITION

Map by ML Design

Library of Congress Cataloging-in-Publication Data has been applied for.

ISBN: 978-0-06-209454-4

11 12 13 14 15 OFF/RRD 10 9 8 7 6 5 4 3 2 1

For Pauline

Contents

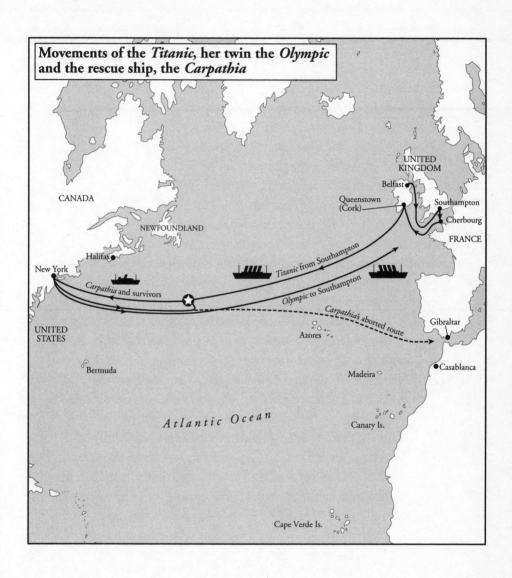

Movements of the *Titanic*, her twin the *Olympic* and the rescue ship, the *Carpathia*

The cover of a 1906 White Star Line passenger list.

J. Bruce Ismay was managing director and chairman of the White Star Line, the company that built the *Titanic*. When the ship struck an iceberg on her maiden voyage, Ismay, who was on board, jumped into one of the last lifeboats to leave. He subsequently became, according to a headline, 'The Most Talked of Man in All the World'. These are some of the things that were said about him:

'Mr Ismay's place as a man and as the responsible director of the White Star Line was on the planks of the imperilled ship. He esteemed his life higher than honour and duty, and as long as this life, which he was so anxious to save, lasts he will bear on his forehead the mark of Cain, the mark of the contempt of all men of honour'
— *Frankfurter Zeitung*

'Mr Ismay cares for nobody but himself. He cares only for his own body, for his own stomach, for his own pride and profit'
— *New York American*

'The humblest emigrant in steerage had more moral right to a seat in the lifeboat than you'
— *John Bull*

'By the supreme artistry of Chance . . . it fell to the lot of that tragic and unhappy gentleman, Mr Bruce Ismay, to be aboard and to be caught by the urgent vacancy in the boat and the snare of the moment'
— H. G. Wells, *Daily Mail*

'You will hunt poor Ismay from court to court, as if he were the only man that was saved'
— G. K. Chesterton, *Illustrated London News*

'I have always felt that he was the most misunderstood and misjudged character of the early part of the century'
— Wilton Oldham, *The Ismay Line*

'The parallel with the tale of Conrad's *Lord Jim* will occur to most of us'
— *New York Tribune*

HOW TO SURVIVE THE *TITANIC*

PART I

At Sea

There was a Ship, quoth he
– Samuel Taylor Coleridge, The Rime of the Ancient Mariner

Chance

I took the chance when it came to me. I did not seek it.

J. Bruce Ismay, *New York World*

Ah! What a chance missed! My God! What a chance missed!

Joseph Conrad, *Lord Jim*

On the night his ship struck the iceberg, J. Bruce Ismay dined in her first-class restaurant with Dr William O'Loughlin, surgeon of the White Star Line for the previous forty years. The two men had shared similar meals on similar crossings, Ismay in his dinner jacket, O'Loughlin in his crisp white uniform. In another part of the dining room a dinner party was taking place in honour of the Captain, E. J. Smith. It was Sunday, 14 April 1912, and the *Titanic,* four days into her maiden voyage, was heading towards New York where she was due to arrive early on Wednesday morning.

After coffee and cigarettes, Ismay retired to his stateroom and was asleep by 11 p.m. He was aware that they were heading into an ice region because at lunchtime that day Captain Smith had handed him a Marconigram from another White Star liner, the *Baltic,* warning of 'icebergs and large quantity of field ice' about 250 miles ahead on the *Titanic*'s course. Ismay had casually slipped the message into his pocket, taking it out later that afternoon to show two passengers, Mrs Marian Thayer and Mrs Emily Ryerson, and handing it back to

Captain Smith shortly before supper so that the warning could be displayed in the officers' chart room. Ismay was not concerned about ice when he turned out his light; it must have been the calmest night ever known on the North Atlantic. The sky was a vault of stars, the sea a sheet of still black, the *Titanic* – the largest moving object on earth – was 46,000 tons of steel and the height of an eleven-storey building. To stand on the deck that night, a passenger later said, 'gave one a sense of wonderful security'.

The collision occurred at 11.40 p.m.; the ship's speed was 22 knots and it took ten seconds for the iceberg to tear a 300-foot gash along her starboard side, slicing open four compartments. The sound, one woman recalled, was like the scraping of a nail along metal; to another it felt as though the ship 'had been seized by a giant hand and shaken once, twice, then stopped dead in its course'. Ismay awoke, his first thought being that the *Titanic* had lost a blade from one of her three propellers. He put on his slippers and padded down the passageway to ask a steward what had happened. The steward did not know, so Ismay returned to his room, put an overcoat and a pair of black evening trousers on top of his pyjamas and, still in his slippered feet, went onto the bridge where Captain Smith told him they had struck a berg. Was the ship damaged? Ismay asked. 'I am afraid she is,' the Captain replied.

The crew were now stirring and a quiet commotion had begun, with stewards knocking on doors to tell the passengers to collect their lifejackets and come up on deck. When Joseph Bell, the Chief Engineer, appeared on the main staircase Ismay asked for his opinion of the damage. Bell said that he thought, or he hoped, that the pumps would control the water for a while. Ismay briefly returned to his room, but he was soon back on the bridge and heard the Captain give the order for lifeboats to be prepared, and for women and children to go first. He then walked along the starboard side of the ship where he met one of the officers and told him to start getting the boats out. It was now five minutes after midnight. *I rendered all the assistance I could*, Ismay later said. *I helped as far as I could.* Staying on the starboard side throughout, he called for women and children to fill Lifeboats 3, 5, 7 and 9 (Lifeboats 2, 4, 6, 8, 10, 12, 14 and 16 were located on the port side). When thirteen of the standard boats were away (2, 4 and 11 had yet to be launched), Ismay helped to load Collapsible C, which was one of the *Titanic's* four Engelhardt canvas-sided life-rafts. Twenty-one women, two men, fourteen young children and six crew were given seats: forty-three passengers so far in a boat which allowed for a maximum of forty-seven. Chief Officer Wilde ordered Collapsible C to be lowered. The deck was flooding and the ship listing heavily. *I was standing by the boat*, Ismay said. *I helped everybody into the boat that was there, and, as the boat was being lowered away, I got in.* He got into the fourteenth boat to be launched and the third-to-last boat to leave the *Titanic* on the starboard side.

———

Ismay later claimed that he had left in the last boat on the starboard side. Other versions of his departure from the ship exist, none of which agree.[1] In the US inquiry, which began in New York's Waldorf-Astoria Hotel five days later, and the British Board of Trade inquiry which opened in London the following month, every man called to the stand had to account for his own survival. Many witnesses were also asked to describe the actions of Ismay: 'Did you see Mr Ismay?', 'What was Mr Ismay doing?' Those – mainly women – who were not invited to give evidence, related their tales of Ismay to the press. Many, like Mrs Malaha Douglas, who was returning home from a furniture-buying trip in Europe with her husband – the heir to Quaker Oats – remembered Ismay getting into the first boat to leave. One of the ship's firemen, Harry Senior, agreed. 'I saw the first boat lowered. Thirteen people were on board, eleven men and two women. Three were millionaires and one was Ismay.' Mrs Charlotte Drake Martinez Cardeza, travelling with fourteen trunks of new clothes from Paris, said that Ismay was the first person to climb into the first boat to leave and that he selected his own crew to row him away. Mrs Cardeza's son Thomas, on the other hand, who was in the same boat as his mother, told a reporter that the women in their boat, one of the last to leave, had begged Ismay to join them. '"Mr Ismay, won't you come with us? We will feel safer." "No," Ismay said, "I will remain here and not take the place of any women."' It was only under pressure, Thomas Cardeza recalled, that Ismay was eventually persuaded to climb in. Edward Brown, a first-class steward, remembered Ismay standing, not on the deck to help the women and children into Collapsible C, but inside the boat itself, ready to receive them. But according to Georgette Magill, aged sixteen, Ismay got into the 'last boat' of all, and only then because he had been ordered to do so by Captain Smith himself.

One reason for the conflicting accounts of Ismay's actions was the chaos of the night. It was, a passenger recalled, 'too kaleidoscopic for me to retain any detailed picture of individual behaviour'.[2] Added to which, very few people, apart from some members of the crew and a small circle of first-class passengers, knew who Ismay was or what he looked like. It was simply not possible to see where amongst the crowds various individuals were standing and into which of the boats they were climbing. Port and starboard were two separate neighbourhoods. The *Titanic* was a sixth of a mile long and the decks were like avenues; the corridors inside were even named after streets such as 'Park Lane' and 'Scotland Road'. It took Second Officer Charles Lightoller, who had spent twenty years at sea, several days before he could find his way from one end of the ship to the other by the shortest route.

Those who did know Ismay had different versions of his departure from the *Titanic*, all of which served their own interests. August Weikman, the ship's barber, swore in an affidavit that Ismay was 'literally thrown' into Collapsible C by an officer: 'Mr Ismay refused to go, when the seaman seized him, rushed him to the rail and hurled him over.' Weikman, who was rescued from a floating deckchair, had been given a first-class barber's shop on every new White Star liner and, as Ismay's personal barber, considered himself a friend. Lightoller, another loyal White Star Line employee, agreed that Ismay had been 'thrown' into a boat by Chief Officer Wilde.

But Ismay always denied the suggestion that he was obeying orders when he jumped into Collapsible C – a fact that would have entirely exonerated him from the accusation of cowardice.

Quartermaster George Rowe, the seaman put in charge of Collapsible C, said under oath that no one invited or ordered Ismay to jump in, and that Ismay had jumped in before, and not after, the boat had started to be lowered. 'The Chief Officer wanted to know if there were any more women and children,' Rowe told the US inquiry. 'There were none in the vicinity. Two gentleman passengers got in; the boat was then lowered.' But forty years later, in a letter to Walter Lord who was compiling eyewitness accounts for his book, *A Night to Remember*, Rowe recalled it differently. In his revised account, he only noticed Ismay's presence when the boat had nearly reached the water level, and he had no idea how or when the owner had left the *Titanic*. 'We had great difficulty in lowering as the ship was well down by the head . . . it was then that I saw Mr Ismay and another gentleman (I think it was a Mr Carter) in the boat.'[3]

William E. Carter, who had jumped into Collapsible C at the same time as Ismay, was an American polo-playing millionaire who belonged to the Philadelphia fast set. The Carters were currently based in England, where their son was at school, and were returning to their country house with a new $5,000 Renault motor car in the hold of the ship. In an interview for *The Times* on 22 April, Carter said that after 'waving' off his wife, Lucille, and their two children who were seated in a lifeboat launched on the port side, he crossed over to starboard where he and Ismay were invited into Collapsible C by an officer: 'As the last boat was being filled we looked around for more women. The women in the boats were mostly steerage passengers. Mr Ismay and myself and several officers walked up and down the deck, crying, "Are there any more women here?" We called for several minutes and got no answer. One of the officers then said that if we wanted to, we could get into the boat if we took the place of seamen. He gave us preference because we were among the first-class passengers. Mr Ismay called again, and after we had no reply we got into the lifeboat. We took oars and rowed.' When their lifeboat was launched, Carter said, 'the deck was deserted'. William Carter corroborates Ismay at every turn: 'our narratives are identical; the circumstances under which we were rescued from the *Titanic* were similar. We left the boat together and were picked up together.'⁴ 'I hope I need not say,' declared Ismay, 'that neither Mr Carter nor myself would, for one moment, have thought of getting into the boat if there had been any women there to go in it.' But Lucille Carter seems to have been still on the ship when her husband jumped into Collapsible C; it was not until fifteen minutes later that she and her children were given places in boat number 4. In 1914 Mrs Carter filed for divorce, claiming among other things that her husband William E. Carter had deserted her on the *Titanic*.

The women passengers in Collapsible C remember the loading of their boat differently. Margaret Devaney, an Irish third-class ticket holder aged nineteen, recalled being 'caught in a crowd and pushed into Collapsible C', and Waika Nakid, a Lebanese passenger also aged nineteen, whose twenty-year-old husband, Sahid, managed to get in with her, saw two men from Lebanon being shot at: terrified for her husband, she and some other women covered Sahid with their clothes. On the twenty-fifth anniversary of the *Titanic*, another Lebanese passenger, Shawneene George, confirmed Waika Nakid's story in an interview with the *Sharon Herald*: 'Sailors armed with revolvers drove the men away from the boats shouting, "women and children first!". They shot into the air to frighten the men. Many passengers were overcome with fright . . . A scared young man leaped over the side of the liner and landed in the bottom of the lifeboat. Women shielded him with their night clothing so that sailors would not see him. They would have shot him.' Reports of gunfire also came from one of the ship's firemen, Walter Hurst, and Hugh Woolner, an English first-class passenger who eventually jumped into Collapsible D (the last boat to be launched) said that 'two flashes of a pistol' alerted him to a group of 'five or six' men climbing into Collapsible C. 'We helped the officer to pull these men out, by their legs and anything we could get hold of.' Woolner then helped to load the boat with women. One of these, Emily Badman, an eighteen-year-old servant from Southampton, told the *Jersey Journal* how she pushed through crowds to get to Collapsible C; May Howard, a twenty-seven-year-old laundry worker emigrating to Canada, told the *Orleans American* that: 'One of the officers grabbed

Mrs Goldsmith and myself and pushed us to the edge of the ship where the lifeboat [Collapsible C] was being filled with women and children first'; Mrs Emily Goldsmith, emigrating to America with her family, said that Collapsible C was surrounded by a line of seamen with linked arms, who were allowing only women and children through. Amy Stanley, a twenty-year-old servant from Oxfordshire, said that 'as we were being lowered a man about 16 Stone jumped [in] almost on top of me. I heard a pistol fired – I believe it was done to frighten the men from rushing the boat.'

A week later, seventeen-year-old Jack Thayer – whose father John B. Thayer, the Second Vice-President of the Philadelphia Railroad, died that night – gave the following account of the activity around Collapsible C in a letter to Judge Charles L. Long, who had lost his son: 'There was an awful crowd around the last boat of the forward part of the starboard side, pushing and shoving wildly.' In a subsequent account, *The Sinking of the SS Titanic*, privately printed for his family in 1940, Jack Thayer recalled hearing an order for 'all women to the port side'. After saying goodbye to his mother, he and his father went over to the starboard side where passengers and crew stood around wondering what was happening. 'It seemed we were always waiting for orders,' he wrote, 'and no orders ever came. No one knew his boat position, as no lifeboat drill had been held.' He then described the scene around Collapsible C: 'There was some disturbance in loading the last two forward starboard boats. A large crowd of men were pressing to get into them. No women were around as far as I could see. I saw Ismay, who had been assisting in the loading of the last boat, push his way into it. It was really every man for himself. Many of the crew and men from the stokehole were lined up, with apparently not a thought of getting into a boat without orders . . . Two men, I think they were dining-room stewards, dropped into the boat from the deck above. As they jumped [an officer] fired twice into the air. I do not believe they were hit, but they were quickly thrown out.'[5] Colonel Archibald Gracie, a first-class passenger, reported that there had been 'no disorder in loading and lowering' Collapsible C. 'Two gentlemen got in, Mr Ismay and Mr Carter. No one told them to get in. No one else was there.'[6]

In an interview with the *New York Times* on 19 April, Abraham Hyman, a thirty-four-year-old framer from Manchester hoping to join his brother in New Jersey, gave his version of the loading of Collapsible C. There was 'so much confusion that nobody knew what was going on . . . some of the people were too excited to understand what was said to them and they crowded forward and then some of the officers came and pushed them back, crying out for women to come first, and some of them said they would shoot any man who tried to get into the boats'. Hyman, whose memory of events comes closest to what must have been the truth, continued:

> We got some of the women and children out of the crush and sent them to where the boats were and saw them get in until I counted on one boat thirty-two persons. Then there was a shout that no more could go into that boat, although I have since heard that the lifeboat could easily hold from forty-five to fifty persons. By this time we all felt sure we would be drowned if we stayed on the [ship] – that is, all of the steerage people thought so. And that was enough to drive them wild and a fight began among them to get to where the boat was being made ready. The forward deck was jammed with the people, all of them pushing and clawing and fighting, and so I walked forward and stepped over the end of the boat that was being got ready [Collapsible C] and sat down.

But in Ismay's account of the sinking of the *Titanic* nothing much happened. There was no crowd around his boat and no panic, despite the fact that most of the lifeboats had now departed, that the boat deck was awash and the ship beginning to list. 'Did you see any struggle among the men to get in?' he was asked at the US Senate inquiry. *None*, he replied.

One after another eighteen lifeboats dropped into the sea, most of them half-filled.

Lifeboat number 1 contained only Lord and Lady Duff Gordon plus their staff and seven crew members. There had not been time to launch Collapsibles A and B, which floated free when the *Titanic* went down and were used as rafts. Because there had been no proper safety drill, most of the crew were unconfident about handling the davits and nervous about filling the boats to capacity in case they buckled. Nor did they know which boat they were assigned to: lists had been posted up, but no one had bothered looking at them. No alarm had been raised, there was no attempt at imposing discipline and no one knew what was happening or what they were meant to be doing. Some boats were manned by stewards who had never held an oar before, and some were rowed by women. Only three of the lifeboats contained lamps, and none contained compasses. Should the lifeboats encounter the 30-mile-wide ice floe that was advancing towards them, sixteen feet above the water level, there was nothing that could be done. Beneath the *Titanic*'s thin gleam of efficiency lay an acreage of slapdash.

The *Titanic* had lifeboat capacity for 1,100 of the 2,340 passengers and crew on board, but only 705 people were saved, of whom 325 were men. In meetings held in October 1909 and January 1910, Ismay – who made the final decisions about the ship's design, decoration and equipment – had turned down the suggestion of placing three boats, rather than one, on each davit. With her sixteen wooden lifeboats and four Engelhardt collapsible boats, the *Titanic* was already carrying 10 per cent more than the British Board of Trade official requirements and anyway, why clutter the recreation deck unnecessarily when the ship was itself a lifeboat? 'If a steamship had enough lifeboats for all,' a White Star Line official patiently explained, 'there would be no room left on deck for the passengers. The necessary number of lifeboats would be carried at the cost of many present comforts to our patrons.' Instead of lifeboats the patrons had luxury: a palm court, a gymnasium and a Louis XVI restaurant. As it was, because the majority of passengers were not told the *Titanic* was sinking and few believed anyway that the ship was sinkable, most thought it safer to stay in the floating palace with its clocks and chairs and electric lights, than commit themselves to an unknown future on the watery wilderness below. 'Most of the men thought they would be safer back on the boat,' Abraham Hyman said, 'and some of them smiled at us as we went down.'

Collapsible C was lowered with difficulty. The *Titanic* was now listing heavily towards starboard, causing the lifeboat to catch on the rivets; Jack Thayer, watching from the deck, 'thought it would never reach the water right side up, but it did'. From inside the boat, Abraham Hyman recalled that 'when we were nearly to the water we passed a big hole in the side of the [ship]. This was about three quarters of the way back toward the stern and the pumps were throwing a great stream of water out through it. It threatened to swamp our boat, and we got scared. There were about ten men in the boat and we each took an oar and pushed the boat away from the side of the ship. That's all that saved us.'

George Rowe, who took charge of Collapsible C, was a thirty-two-year-old former Merchant Marine from Hampshire. Apart from Rowe, Ismay, William Carter, Albert Pearcey, Margaret Devaney, Emily Badman, May Howard, Amy Stanley, Emily Goldsmith, Shawneene George, the Nakids, and Abraham Hyman, whose experiences of the loading we have already heard, the boat contained three firemen and thirty-one further adults and children (all third-class), including twenty-three of the *Titanic*'s seventy-nine Lebanese passengers (of whom thirty-one were saved altogether). 'The boat would have accommodated certainly six more passengers,' Ismay said in his public statement to the press, 'if there had been any on the boat deck to go.'

The *Titanic* was only superficially a liner for the rich: she was actually an emigrant ship. Ismay's lifeboat consisted of the following people, several of whom were returning from visits to their family while others were hoping to start new lives:

Mrs Mariana Assaf, aged forty-five, living in Canada
Mrs Mary Abrahim, aged eighteen, living in Pennsylvania
Mrs Latifa Baclini and her three daughters: Marie, aged five, Eugenie, aged three, and Helene, nine months

Mrs Catherine Joseph, aged twenty-four, returning to her husband in Detroit with her four-year-old son, Michael (who was placed in another lifeboat), and her two-year-old daughter, Anna

Mrs Darwis Touma, aged twenty-three, returning to her husband in Michigan with her children, Maria and Georges

Mrs Omine Moubarek, emigrating to Wilkes-Barre, Pennsylvania, with her two sons, Gerios, aged seven and Halim, aged four

Banoura Ayorb Daher, aged fifteen, emigrating to Canada

Adele Najib Kiamie, aged fifteen, travelling from Lebanon to be married

Fatima Mousselmani, aged twenty-two, travelling from Lebanon to be married

Jamilia Nicola-Yarred, aged fourteen, and her eleven-year-old brother, Elias, travelling to meet their father

Hilda Hellström, aged twenty-two , travelling from Sweden

Velin Öhman, aged twenty-two, travelling from Sweden

Sarah Roth, aged twenty-six, a tailor from Southampton

Anna Salkjelsvik, aged twenty-one, travelling from Norway

John Goldsmith, aged ten

Lee Bing, Chang Chip, Ling Hee, Ali Lam: Chinese employees of the Donaldson Line. Thought by Ismay to be stowaways, they had boarded the ship, along with four other men, on one third-class ticket.

At forty-nine, Ismay was the oldest person there and the oldest man by ten years. Six foot four with a waxed moustache and the handsome face of a matinee idol, he sat amongst Lebanese, Chinese and Swedish passengers in his pyjamas, coat and slippers and spoke to no one. It is unlikely that anyone, apart from William Carter, had any idea who he was.

The *Titanic* went down two hours and forty minutes after she hit the iceberg – the same length of time, it has been noted, as a performance in the theatre. For one passenger at least, the events of the night seemed 'like a play, like a drama that was being enacted for entertainment', and for Ismay the drama would continue before a variety of audiences and on a number of different stages.

In her bowels the ship carried 3,500 mailbags containing 200,000 letters and packages. One of them was the manuscript of a story by Joseph Conrad called 'Karain: A Memory', which he was sending to New York. 'Karain' is the tale of a man who impulsively betrays a code of honour and lives on under the strain of intolerable guilt.

———

Ismay said he was pushing an oar with his back to the ship when the *Titanic* made her final plunge. *I did not wish to see her go down,* he told the US inquiry, adding, *I am glad I did not.* He was one of few survivors not to have looked. 'I will never forget the terrible beauty of the *Titanic* at that moment,' wrote Charlotte Collyer, in anticipation of Yeats's phrase in 'Easter 1916'. 'She was tilted forward, head down, with her first funnel partly under water. To me she looked like an enormous glow worm: for she was alight from the rising water line, clear to her stern – electric lights blazing in every cabin, lights on all the decks and lights at her mast heads.'[7] The glow worm carried Charlotte Collyer's young husband and all the money and possessions they had in the world. 'Fascinated', wrote Jack Thayer, who had jumped into the sea and was swimming away. 'I seemed tied to the spot. Already I was tired out with the cold and struggling, although the life preserver held me head and shoulders above water.' 'Fascinated', wrote Elizabeth Shutes, a governess travelling first-class with her charge, 'I watched that black outline until the end.'[8] 'Fascinated', repeated Violet Jessop, one of the ship's stewardesses, who was in a lifeboat; 'my eyes never left the ship, as if by looking I could keep her afloat . . . I sat paralysed with cold and misery as I watched *Titanic* give a lurch forward. One of the huge funnels toppled off like a cardboard model, falling into the sea with a fearful roar. A few cries came to us across the water, then silence, as the ship seemed to right herself like a hurt animal with a broken back.'[9] The *Titanic* went down 'like a stricken animal', agreed Lawrence Beesley, a second-class passenger travelling to a Christian Science gathering. 'We had no eyes for anything but the ship we had just left. As the oarsmen pulled slowly away we all turned and took a long look at the mighty vessel towering high above our midget boat, and I know it must have been the most extraordinary sight I shall ever be called upon to witness.' With her lights ablaze, the *Titanic* resembled, until her final plunge, a living, breathing thing.

When the *Titanic* had melted away, those on the moonless water gazed with disbelief at the great ship's disappearance. She was, they now realised, a boat and not a palace; it is as though they had expected to see her crumble to the ground. In the midst of horror ('the horror, the helpless horror', said Elizabeth Shutes, echoing the famous last words of Kurtz in Joseph Conrad's *Heart of Darkness*) they found themselves caught in contemplation. Second Officer Lightoller, who was swept overboard and blown across the sea by a rush of steam towards an upside-down lifeboat onto which he clung and which was itself thrown from the site of the ship by a falling funnel, described watching the 'terrible awe-inspiring sight' of 'this unparalleled tragedy that was being enacted before our very eyes . . . the huge ship slowly but surely reared herself on end and brought rudder and propellers clear of the water, till at last, she assumed an *absolute perpendicular position*. In this amazing attitude she remained for the space of half a minute. Then with impressive majesty and ever-increasing momentum, she silently took her last tragic dive to seek a final resting place in the unfathomable depths of the cold grey Atlantic . . .' 'I realise', Lawrence Beesley said, 'how totally inadequate language is to convey to some other person who was not there any real impression of what we saw . . .'[10] But those who watched, from the unlettered to the urban sophisticates, each described the same experience of romantic awe and dread fascination. The sinking of the *Titanic* seemed to Edith Russell, a thirty-four-year-old first-class passenger who worked as a fashion buyer and a reporter for *Women's Wear Daily*, like the collapse of a 'skyscraper'. But in her death throes, the *Titanic* also became Excalibur, the sword of the mortally wounded King Arthur which was thrown back into the lake and caught by a mysterious hand. Her glittering scabbard sparkled and flashed one last time before the surface of the water closed over her, erasing all trace.

The *Titanic* left behind a level sea, undisturbed apart from a reddish stain and a tangle of flotsam consisting mainly of deck-chairs, the cork of lifejackets and the bodies of the dead and dying. Hanging over the water, 'like a pall', was what Colonel Archibald Gracie – an American first-class passenger saved by climbing onto the upturned Collapsible B – described as 'a thin, light-grey smoky vapour' which 'produced a supernatural effect, and the pictures I had seen by Dante and the description I had read in my Virgil of the infernal regions, of Charon, and the River Lethe, were then upper-most in my thoughts'.[11]

While many of those in the lifeboats had not realised until the very end that the *Titanic* would sink, only Ismay and the surviving officers were prepared for what came next. There rose from the wreckage what Colonel Gracie called 'the most horrible sounds ever heard. The agonising cries of death from over a thousand throats, the wails and groans of the suffering, the shrieks of the terror-stricken and the awful gaspings for breath of those in the last throes of drowning, none of us will ever forget to our dying day.' In his account of the birth and death of the *Titanic*, the Irish journalist Filson Young had described the noise of the Belfast shipyard at Harland & Wolff when the 'monster' was under construction as 'a low sonorous murmuring like the sound of bees in a giant hive'. In his memoirs, Second Officer Lightoller described the *Titanic* while she was being prepared as 'a nest of bees' which became on sailing day 'a hive about to swarm'.[12] With the ship's disappearance the sound returned. 'The *Titanic* was like a swarming bee-hive,' recalled Charlotte Collyer, 'but the bees were men . . . Cries more terrible than I had ever heard rang in my ears.'

'We could see groups of almost fifteen hundred people still aboard,' recalled Jack Thayer, 'clinging in clusters or bunches, like swarming bees.' The image recalls the journey made by Aeneas to the underworld in Dante's *Inferno*, where he passes a valley in which he hears what he thinks are bees in a summer meadow. The humming,

Aeneas is told, comes not from bees but from the souls of the dead drinking from the River Lethe, in order to forget their previous lives before returning to live for a second time on land. The cries also recall the passage in Rabelais's *Gargantua and Pantagruel*, where the feasting passengers aboard a ship are terrified to hear the voices of people talking in the air. 'Let us fly,' they say, 'Larboard! Starboard! Topsails! Scudding-sails! We're all dead men. Let us fly, in the name of all devils, fly!' Pantagruel's ship was reaching 'the approaches of a frozen sea over which there was a huge and cruel combat at the onset of last winter'; the words they heard were the cries of men and women which had frozen in the air.[13]

Most of the survivors had not known until now that there were people still aboard the *Titanic*, that the ship was not equipped to save everyone. Edith Russell recalled that moments before the *Titanic* went down 'there was a loud cry, as if emanating from one throat. The men in our boat asked the women to cheer, saying "those cheers that you hear in the big boat mean they have all gotten into lifeboats and are saved". And do you know, that we actually cheered, believing that the big shout was one of thanksgiving.'[14] In a letter sent forty-five years later to Walter Lord, Lawrence Beesley described how they 'had no knowledge of the number of boats available, how many had got away in the boats, and, in fact, no reason to doubt that every one on board had, in some way, a chance of safety. There was', he said, 'no forewarning of the tragedy, no anticipation of peril for our fellow passengers as we saw the *Titanic* glide slowly down to her doom. The ship had gone but the passengers were, in all probability, safe, waiting only as we did, for the dawn to bring the rescuing ships. Therefore the terrible nature of the cries, which reached us almost immediately after the *Titanic* sunk, came upon us entirely unprepared for their terrible message. They came as a thunderbolt, unexpected, inconceivable, incredible. No one in any of the boats standing off a few hundred yards away can have escaped the paralysing shock of knowing that so short a distance away a tragedy, unbelievable in its magnitude, was being enacted, which we, helpless, could in no way avert or diminish.'[15]

The lifeboats might have returned to rescue some of the many hundreds of people in the water, most of whom, floating in their lifejackets, would die of hypothermia, but both passengers and crew refused, fearing that suction from the sinking ship would pull them down, if they were not overwhelmed by soaking bodies trying to clamber to safety. To return would be suicide, they reasoned. In Lifeboat 1, referred to afterwards as the Money Boat, Lady Duff Gordon, also known as the fashion designer Lucille, watched the ship disappear and then commiserated with her maid: 'You have lost your beautiful night-dress.' Lord Duff Gordon was reported to have offered the seven crew members who were rowing them away £5 each for a new uniform if they did not turn back. Jack Thayer, struggling in the water, watched while the lifeboats 'four or five hundred yards away, listen[ed] to the cries and still they did not come back. If they had turned back several hundred more would have been saved. No one can explain it. It was not satisfactorily explained in any investigation. It was just one of the many "Acts of God" running through the whole disaster.'

The only boat to attempt further rescue was Lifeboat 14, in the charge of Fifth Officer Lowe. Instead of taking his cargo into the swarm of struggling bodies, Lowe kept close to the wreckage and held himself 'ready to save any poor man who might become detached from the big confused mass'. He rescued four people, including a Japanese passenger who had strapped himself to a door and seemed as lifeless as a marble statue. 'What's the use,' Lowe had initially said. 'He's dead, likely, and if he isn't there's others better worth saving than a Jap.' But once the rescued man recovered he jumped up, stretched, stamped his feet and took over an oar. 'By Jove,' exclaimed Lowe. 'I'm ashamed of what I said about the little blighter. I'd save the likes o' him six times over if I got the chance.'

Returning to save any of the gasping bodies was not discussed in Collapsible C. Instead, they looked for other boats. As Ismay's was one of the three boats equipped with a lamp he could at least see, if he wished, the passengers who had paid for their tickets on the *Titanic* now floating in the water. But he did not look; nor did he turn his head. Ismay kept his eyes blank and fixed at some point in the distance.

"All the News That's Fit to Print."

The New York Times.

THE WEATHER.

VOL. LXI...NO. 19,808.   NEW YORK, TUESDAY, APRIL 16, 1912.—TWENTY-FOUR PAGES.   ONE CENT

TITANIC SINKS FOUR HOURS AFTER HITTING ICEBERG; 866 RESCUED BY CARPATHIA, PROBABLY 1250 PERISH; ISMAY SAFE, MRS. ASTOR MAYBE, NOTED NAMES MISSING

Col. Astor and Bride, Isidor Straus and Wife, and Maj. Butt Aboard.

"RULE OF SEA" FOLLOWED

Women and Children Put Over in Lifeboats and Are Supposed to be Safe on Carpathia.

PICKED UP AFTER 8 HOURS

Vincent Astor Calls at White Star Office for News of His Father and Leaves Weeping.

FRANKLIN HOPEFUL ALL DAY

Manager of the Line Insisted Titanic Was Unsinkable Even After She Had Gone Down.

HEAD OF THE LINE ABOARD

J. Bruce Ismay Making First Trip to Algonta Ship That Was to Surpass All Others.

Biggest Liner Plunges to the Bottom at 2:20 A. M.

RESCUERS THERE TOO LATE

Except to Pick Up the Few Hundreds Who Took to the Lifeboats.

WOMEN AND CHILDREN FIRST

Cunarder Carpathia Rushing to New York with the Survivors.

SEA SEARCH FOR OTHERS

The California Stands By on Chance of Picking Up Other Boats or Rafts.

OLYMPIC SENDS THE NEWS

By Tuesday 16 April, the whole world knew who J. Bruce Ismay was. The *New York Times* ran the headline: 'Probably 1250 perish; Ismay safe'; the *Denver Post* was amongst dozens of papers to ask: 'Who would not rather die a hero than live a coward?' and the *New York American* summed up the case: 'Mr Ismay cares for nobody but himself. He cares only for his own body, for his own stomach, for his own pride and profit. He passes through the most stupendous tragedy untouched and unmoved. He leaves his ship to sink with its powerless cargo of lives and does not care to lift his eyes. He crawls through unspeakable disgrace to his own safety.' In England, Horatio Bottomley, spokesman for the working man and owner of the weekly, *John Bull*, addressed Ismay directly: 'You were the one person on board who, as chairman of the White Star Line, had a large pecuniary interest in the voyage, and your place was at the Captain's side till every man, woman and child was safely off the ship. The humblest emigrant in steerage had more moral right to a seat in the lifeboat than you.' *John Bull* ran a three-month campaign against him, consisting of telegrams, articles, cartoons, letters and doggerel verses. 'Someone', wrote Bottomley, 'ought to hang over this business.'

Sixteen hundred men and women have been murdered on high seas.'
In the *Daily Mail,* H. G. Wells, the literary giant of the age, was
struck by the role played by chance in Ismay's escape. 'By the supreme
artistry of Chance . . . it fell to the lot of that tragic and unhappy
gentleman . . . to be aboard and to be caught by the urgent vacancy
in the boat and the snare of the moment.' Let 'no untried man', Wells
continued, 'say he would have behaved better in his place. But for
capitalism and for our existing social system [Ismay's] escape – with
five and fifty third-class children waiting below to drown – was the
abandonment of every noble pretension. It is not the man I would
criticise, but the manifest absence of any such sense of the supreme
dignity of his position as would have sustained him in that crisis. He
was a rich man and a ruling man, but in the test he was not a proud
man.'

The most serious accusation against Ismay was that he put profit
before lives, dictating the ship's speed to the Captain in order to get
to New York in record time. 'The *Titanic* disaster is a capitalist disaster,' a radical Kansas paper declared, and 'Ismay, worth one hundred
million dollars and drawing a salary from the White Star Line of one
hundred and seventy five thousand dollars' is 'the epitome of capitalism'. But the most common accusation was that Ismay had not
behaved as a gentleman. 'Gentlemen', wrote Filson Young, 'simply
stood about the decks, smoking cigarettes, talking to one another,
and waiting for the hour to strike. There is nothing so entirely dignified, as to be silent and quiet in the face of an approaching horror.'
Ismay's actions were compared to those of other millionaires, men
like Benjamin Guggenheim who had put on his dinner jacket and
reputedly told his steward: 'We're dressed in our best and are prepared
to go down like gentlemen. I am willing to remain and play the man's
game if there are not enough boats for more than the women and
children.' The night the *Titanic* went down, 'all the great virtues of
the soul were displayed', wrote Philip Gibbs in a forty-page account,
'The Deathless Story of the *Titanic*', published two weeks after the
wreck in a special edition of the mass circulation British paper, *Lloyd's
Weekly News*: 'courage, self-forgetfulness, self-sacrifice, love, devotion

to the highest ideals'. Logan Marshall, whose book, *Sinking of the Titanic and Great Sea Disasters* also appeared in 1912, suggested that the term 'chivalry' was 'a mild appellation' for the 'conduct of the gentlemen' on board. 'Some of the vaunted knights of old were desperate cowards by comparison. A fight in the open field, or jousting in the tournament, did not call out manhood in a man as did the waiting till the great ship took the final plunge.' In the *Daily News*, George Bernard Shaw railed against the romantic language of heroism, the 'ghastly, blasphemous, inhuman, braggartly lying' in which the catastrophe was couched. There was, for Shaw, 'no heroism in being drowned when you cannot help it'.

Social, rather than evolutionary, order must be maintained in a shipwreck: the fittest are obliged to ensure the survival of the weakest. Two days after the *Titanic* went down, the *St Louis Post-Dispatch* put the dilemma for male passengers in the form of an illustration entitled 'The Last Seat – Should He Take It?' A handsome young bridegroom is poised on a ladder between a sinking ship and a departing lifeboat, his lovely wife stands in the lifeboat with her arms open wide; behind him on the ladder is an unknown damsel in distress. There is only one seat in the boat: should the bridegroom preserve the family and join his wife, or be chivalric and make his wife a widow? Ismay, who had no relatives on board the ship to whom he could display his courage, showed neither the manners of a gentleman nor the mettle of a man; like Lady Macbeth, he was unsexed by his crime.

It is easy to become what Second Officer Lightoller called an 'armchair judge', and forget that the *Titanic* passengers were responding to the crisis moment by moment. People behaved instinctively, but instincts need to be trained. Any number of brave and extraordinary acts took place as fear set in, couples separated and people prepared to die, but Ismay was seen to perform none of them. He had one view only: to carry on living. His instinct to survive became a force which overwhelmed and obliterated a lifetime of training. Of the few passengers who publicly praised his behaviour, three were women. The first was a stewardess who said that Ismay had saved her life by insisting that, being a 'woman' as well as a stewardess, she climb into a lifeboat; the second was Edith Russell, the fashion reporter who had likened the ship going down to a skyscraper. Miss Russell told the *New York Times* that as one of the last boats was being lowered, Ismay seized her arm and said: 'Woman, what are you doing here? All women should be off the boat.' Having been told by an officer that there was 'no immediate danger' and that the *Titanic*'s sister, the *Olympic,* was on her way to pick up the remaining passengers, Edith Russell had been drifting calmly between the lounge and the boat deck cradling her 'lucky pig'. Had Ismay not thrown her down the steps to A Deck, where she was shoved headfirst, in her hobble skirt and diamond-buckled slippers, through a porthole and into a lifeboat, she would doubtless have stayed on board. 'There has been much criticism of Mr Ismay,' said Miss Russell, who claimed that they had subsequently become lovers, 'but he certainly saved my life.' The third woman to defend Ismay was Mrs Marian Thayer, who let the press know that any unfair statements about him which had appeared in the papers under her name had not been authorised: 'Mr Ismay had done everything a man could do to help passengers on the *Titanic*.' And Mrs Thayer did everything she could to defend Mr Ismay, who had – although she did not yet know it – fallen in love with her during the voyage.

———

J. Bruce Ismay died on the night of 14–15 April 1912, and died again in his bedroom twenty-five years later. He was mired in the moment of his jump; his life was defined by a decision he made in an instant. Other survivors of the *Titanic* were able, in varying degrees, to pick themselves up and move on, but Ismay was not. His was now a posthumous existence.

It is typical that in his most significant gesture Ismay is poised iconically between a sinking ship and a collapsible boat. The threshold was his natural home. He lived his life between the Old World and the New; when he was in New York, he referred to England as 'the other side'. He crossed the North Atlantic more than eighty times: he held the record number of crossings between England and America, spending the equivalent of two and a half years floating between ports. Whether dining with the doctor on board as though he were an honorary member of the crew, or sitting stiffly amongst the Lebanese women and children in the lifeboat as though he were a passenger, Ismay is always awkwardly positioned, in neither one place nor another. He is never at the centre of his own story. We watch him skulk around the margins of films like *A Night to Remember* and James Cameron's *Titanic*, the image of wealth, weakness and self-interest, an almost sickly figure separated from the healthy fellowship of potential heroes, an idler amongst workers who gets in everyone's way when he is only trying to help. In contrast with the image of the stalwart crew, the back-slapping young men and the blue-blooded passengers, Ismay is portrayed as introverted, impotent, nouveau riche. He is the man who doesn't understand the jokes and can't keep up with the conversation; in Cameron's *Titanic*, where he is played by Jonathan Hyde, Ismay is asked whether it was he who thought of the ship's name.

'Yes, actually,' Ismay replies. 'I want to convey sheer size; and size means stability, luxury and, above all, strength.'

'Do you know of Dr Freud, Mr Ismay?' says Rose De Witt Bukater. 'His ideas about the male preoccupation with size might be of particular interest to you.'

'Freud?' says Ismay. 'Who is he? A passenger?'

Even before the *Titanic*, Ismay was despised in America for inheriting rather than earning his position in the White Star Line. To a country of self-made men, he was a spoilt product of the class system. He was despised in Britain, on the other hand, for his lack of pedigree, for assuming the privileges of an aristocrat while working in trade for a living. For a country whose hierarchy is based on birth, Ismay was no more than a Liverpudlian businessman. Seamen thought him a shore-dweller; to his family, who feared him, he was king of an ocean world. Ismay's problem, as the British inquiry concluded, was that he belonged in a category of his own. He fitted in nowhere, least of all on board the *Titanic* where, when the crisis

came, he slipped through the all-important classifications. Alone of
those on board, his journey did not have a destination. Other passengers were going somewhere, most of them on a one-way ticket:
emigrating, joining their families, getting married, returning home.
But Ismay was drifting across the Atlantic simply to drift back again
when the *Titanic* turned around.

When it came to saving his own life, he later told journalists: 'I took
my chance to escape – yes. It came to me though, I did not seek
it . . . And why shouldn't I take my turn? There are only two classes on
a ship – crew and passengers. I was a passenger. It is true I am president
of the company, but where do you draw the line?'[16] Because he was a
'passenger' and not a member of the crew, Ismay said that he had
'nothing to do with' ordering the loading of the boats. But in a private
account of his own experience, Karl H. Behr, a twenty-six-year-old
love-sick tennis player who, having secretly pursued his sweetheart,
Helen Newsome, across Europe where she was travelling with her
parents, had planned an 'accidental' meeting with her on board the
Titanic, described how Ismay commanded the lowering of the lifeboats
as though he were the most authoritative person on the ship. Behr was
standing with Miss Newsome and her mother and stepfather, Mr and
Mrs Beckwith, next to a 'comfortably filled' lifeboat, when 'Ismay came
over to us and calmly told us we should get into a second boat which
was being filled'. No one obeyed him as he walked off, because they did
not believe that the boat was capable of sinking. 'In a few minutes
Ismay noticed us still standing together; he again walked over and with
considerably more emphasis told us to get into the lifeboat – we were
the last passengers on the deck. I told Mrs Beckwith I thought we
should do what he said, and she finally led the way to the boat. Stepping
in front of Ismay, she asked if all her party could get into the same lifeboat and he replied, "Of course Madam, every one of you".'[17]

———

The first *Titanic* film, *Saved from the Titanic,* appeared after four
weeks, in May 1912, with the twenty-two-year-old silent movie star,

Dorothy Gibson, who had been on board, playing herself in the dress she had worn in her lifeboat. When, in 1954, Walter Lord was writing his classic account of the wreck, *A Night to Remember*, he received hundreds of letters in response to his request for memories. In 1958, when the book was being filmed, dozens more *Titanic* survivors offered the producer, William Macquitty, their help, all believing that they had the 'correct' version of events. Many said they had been among the last to leave the ship, and a few proudly claimed to have been the notorious male who had dressed in women's clothing to ensure his escape. The need felt by the survivors to tell their tales was, from the start, overwhelming and the need of those who were not on board to read their accounts, to see the films, to repeat the experience and work it through, to raise the *Titanic* and watch her go down again and again is one of the shipwreck's most peculiar effects. Colonel Archibald Gracie said, after he was rescued by the *Carpathia*, that the only purpose of the rest of his life was to write an account of surviving the *Titanic*; when he died eight months later he had completed his task.

Sea stories are as fuelled by jumps as romances are by misunderstandings: whether it is jumping off or onto a ship, a jump contains a concentration of narrative intensity. In Ismay's jump can be seen his whole life story, but after the inquiries, while passengers and journalists were putting together the most enduring sea story of the last hundred years, he never spoke of his beloved ship again, either in public or in private. And yet from the age of twenty until his death, he scoured the newspapers of the world, cutting out and keeping every article – and there are thousands – referring either to himself or to the White Star Line. Noting in his square, slow handwriting the source and date of each, he pasted them in chronological order into a dozen large, labelled leather-bound ledgers. In this sense, Ismay compiled an edited version of his rise and fall.

Despite his silence, he is someone we already know. Ismay has long inhabited the seabed of our psyches; he has been wandering through our literature for centuries in his various guises. He is Noah, building his ark into which everything comes in pairs; he is

Moby Dick's Captain Ahab, whose obsession with the great white Leviathan dooms his crew; he is Ishmael – even the name is similar – who 'alone survived' when the *Pequod* went down. Ismay is the Ancient Mariner, cast out from the community of mankind; he is Dr Frankenstein, whose creation became a monster that pursued him across icy wastes; and he is Joseph Conrad's Lord Jim, who also jumped from a sinking ship and was forced to live on without honour.

Luckless Yamsi

There is something peculiar in a small boat upon the wide sea. Over the lives borne from under the shadow of death there seems to fall the shadow of madness.

Joseph Conrad, *Lord Jim*

The small boats drifted through the ice. Above them hung a sky of spectacular brilliance: what looked at first like the lights of ships turned out to be falling stars whose reflections shot across the water like cat's eyes. With the morning sun icebergs the size of islands became glittering gems; some were shades of pink and blue, others were gleaming pyramids of beaten gold. Their 'awful beauty', wrote Lawrence Beesley, 'could not be overlooked'. It was only now that those inside the boats were able to see one another's faces. A spirited American first-class passenger, who became known as the 'unsinkable' Molly Brown, described the morning of 15 April as the 'most wonderful' she had ever seen. 'I have just returned from Egypt. I have been all over the world, but I have never seen anything like this. First the gray and then the flood of light. Then the sun came up in a ball of red fire. For the first time we saw where we were. Near us was open water, but on every side was ice. Ice ten feet high was everywhere, and to the right and left and back and front were icebergs. Some of them were mountain-high. This sea of ice was forty miles wide, they told me.' Ismay did not remember the sunrise that day.

When the distress call from the *Titanic* came in, Harold Cottam, the twenty-one-year-old wireless officer on the Cunard Line's *Carpathia*, bound from New York to Gibraltar, was preparing for bed but happened to still have the telephone to his ear. Had the message arrived a few minutes later, the Marconi machine would have been turned off. Cottam informed the Captain, Arthur Rostron, who headed to the spot where the *Titanic* was reported wounded. Rostron had no idea how many passengers he was to pick up, how many other ships would be on the scene, or in what state he would find the great liner. As the *Carpathia* steamed ahead he prepared a list of orders for his crew:

English doctor, with assistants, to remain in first-class dining room.

Italian doctor, with assistants, to remain in second-class dining room.

Hungarian doctor, with assistants, to remain in third-class dining room.

Each doctor to have supplies of restoratives, stimulants, and everything to hand for immediate needs of probable wounded or sick.

Purser, with assistant purser and chief-steward, to receive the passengers etc., at different gangways, controlling our own stewards in assisting *Titanic* passengers to the dining rooms, etc.; also to get Christian and surnames of all survivors as soon as possible to send by wireless.

Inspector, steerage stewards, and Master at Arms to control our own steerage passengers and keep them out of the third-class dining hall, and also to keep them out of the way and off the deck to prevent confusion.

Chief Steward: that all hands would be called and to have coffee, tea, soup, etc., in each saloon, blankets in saloons, at the gangways, and some for the boats.

To see all rescued cared for and immediate wants attended to.

My cabin and all officials' cabins to be given up. Smoke rooms, library etc. dining rooms, would be utilised for *Titanic*'s passengers, and get all our own steerage passengers grouped together. [1]

When the *Carpathia* arrived at 4.30 a.m. there was nothing to see but 'boxes and coats and what looked like oil on the water'.[2] The lifeboats, scattered across a five-mile radius, slowly gathered around the rescue ship and at six that morning Ismay was picked up. He said that he had been rowing continuously, but his emotional state when he left the *Titanic* suggests that he was incapable of doing anything physical, while his frozen condition when the *Carpathia* arrived implies that he had not moved a muscle for hours. In an interview with the *Guernsey Press*, a *Titanic* first-class stewardess called Annie Martin said that she recalled Ismay 'sitting on his haunches on the stern of the boat that was cleared by the *Carpathia* just before ours. He sat there like a statue, blue with cold, and neither said a word nor looked at us. He was nearly dead when taken on board, for he was wearing only his nightclothes and an overcoat.'

The 700 *Carpathia* passengers, mostly American tourists taking their spring break, watched from the rails as the *Titanic* survivors were hoisted up. Ropes were tied to the waists of the adults to support them as they climbed the Jacob's ladder, the sick and wounded were lifted up on a makeshift chair swing, and the babies and small children were carried in canvas ash bags. Captain Rostron noted the extraordinary silence of them all, the marked absence of any excitement or response. The *Carpathia* was soon joined by the *Californian*, which had stopped her engines in the midst of an ice field the night before, eight miles from where the *Titanic* would sink. When the *Californian* had sent an ice warning earlier in the evening, Jack Philips, one of the *Titanic*'s two overworked wireless operators, told them to 'shut up' and 'keep out' as he had urgent messages to deal with, most of them cheerful Marconigrams which the passengers were sending home. The *Californian*'s disgruntled wireless operator therefore took off his head-phones and went to sleep, and the crew on the night watch wondered why the *Titanic* was firing rockets at regular intervals.

At 8.30 a.m., when the lifeboats had all been accounted for, there was a roll call and the names of the missing became known. The expressions of grief were class-dependant: the women in first-class held themselves together with decorum while those in steerage wept.

'When it seemed sure that we should not find any more persons alive,' one of the *Carpathia*'s stewards told the press, 'the bedlam came. I hope never to go through it again. The way those women took on for the folk they had lost was awful. We could not do anything to quiet them until they had cried themselves out.'³ Captain Rostron turned the *Carpathia* back in the direction of New York while the *Californian* stayed behind to search for bodies. One survivor told a reporter that: 'While we were on the *Carpathia* we passed through a school of about a dozen whales and later on we passed a seal that was floating on a cake of ice. A little farther on we passed a big floe of ice on which there was a big white polar bear prowling around.'⁴

Ismay's first action on board, according to one of the ship's officers, was to take himself to the dining room and announce 'I'm Ismay, for God's sake get me something to eat.' Having eaten, he then reportedly tried to pay the steward $2 before demanding a stateroom in which to rest. Mrs Lucien P. Smith, one of eleven brides to be widowed on her honeymoon, described him stepping onto the *Carpathia* shouting: 'I'm Ismay! I'm Ismay! Get me a stateroom!' 'I know many women who slept on the floor in the smoking room,' Mrs Smith said, 'while Mr Ismay occupied the best room . . . being in the centre of the boat, with every attention, and a sign on the door, "Please do not knock".' But far from drawing attention to himself, Ismay wanted to get away from the others as quickly as possible. As hellish to deal with as the sinking of his ship was the expectation that he blend in with the brotherhood of *Titanic* survivors on the *Carpathia*. He disappeared long before the 8.30 roll call would have alerted anyone to his presence. At 7.30 that morning, Captain Haddock, commander of the *Titanic*'s twin sister the *Olympic*, currently also crossing the North Atlantic, received a Marconigram from Captain Rostron informing him that 'Mr Bruce Ismay is under an opiate'.

In a statement given to the US Senate inquiry, the longest and most detailed account of his actions he would make, Ismay gave his own version of what happened when he boarded the *Carpathia*.

I understand that my behaviour on board the *Titanic* and subsequently
on board the *Carpathia* has been very severely criticised . . . So far as
the *Carpathia* is concerned, Sir, when I got on board the ship I stood
with my back against the bulkhead, and somebody came up to me
and said, 'Will you not go into the saloon and get some soup or
something to drink?' 'No,' I said, 'I really do not want anything at
all.' He said, 'Do go and get something.' I said, 'No. If you will get
me in some room where I can be quiet, I wish you would.' He said,
'Please go into the saloon and get something hot.' I said, ' I would
rather not.' Then he took me and put me into a room. I did not know
whose room the room was, at all. The man proved to be the doctor of
the *Carpathia*. I was in that room until I left the ship. I was never
outside the door of that room. During the whole of the time I was in
this room, I never had anything of a solid nature at all; I lived on
soup. I did not want very much of anything. The room was constantly
being entered by people asking for the doctor. The doctor did not
sleep in the room the first night. The doctor slept in the room the
other nights that I was on board that ship.[5]

His defence of the privileged treatment he received – that his
small room had not been sufficiently private and that he was continually interrupted – did little to soften the mood of the American
public. 'Ismay crawls through unspeakable disgrace to his own safety,'
wrote the *New York American* when the news reached land that the
owner of the ship had survived. He 'seizes hold of the best accommodation in the *Carpathia* to hold communion with his own
unapproachable conduct'. Except that Ismay did not see his conduct
as unapproachable. He did not feel, as he lay in his stateroom, that
he had betrayed his better self or that a hairline crack in his character
had appeared when he jumped into the lifeboat: *As I lay in my stateroom on board the Carpathia,* he said in the later inquiry, *I went over
every detail of the affair. There was nothing that I did that I am sorry for.
I can truthfully say that my conscience is clear.*[6]

Ismay cut himself off from company at the best of times; it was his
instinct to escape from a crowd. His father, who founded the White

Star Line, had been a congenial man who, when on board his own ships, spent his days with the Captain on the bridge and his evenings with the better class of passenger. But when Ismay crossed the Atlantic he preferred to be alone in his cabin looking over papers and reports. He felt no common ties; he was disconnected from whatever it was that bound people to one another. On the *Carpathia* he did not ensure that a list of the surviving crew was sent to their increasingly desperate families in Southampton (90 per cent of the ship's crew were born or lived in the port), or inquire after the comfort and welfare of the surviving passengers. In an article in the *National Magazine* which appeared the following October, Marie Young, a first-class American passenger, described how 'the President of the White Star Line – hidden in the English physician's comfortable room . . . voyaged to New York as heedlessly indifferent to the discomfort of his Company's passengers as he had been to the deadly peril that had menaced them'. On the *Titanic*, Ismay was the only man, apart from the Captain, the ship's designer, Thomas Andrews, and the Chief Engineer, Joseph Bell, who knew that the ship would be at the bottom of the sea within two hours and that the available lifeboats could carry less than half of those on board, but he did not warn, or even try to find, Dr O'Loughlin, who had been his dining companion that evening, or Charles M. Hays, who was travelling as his guest, or Richard Fry, his valet of ten years, or George Dodd, his family's former butler, now working as a steward in first-class, or Arthur Hayter, who had been his steward throughout the journey, or William Harrison, his secretary, all of whom were to die ('How my husband loved his work,' Harrison's grieving widow later wrote to Ismay; 'he was so proud of his position as a private help to you.'[7]) Other people held no reality for Ismay. It was only Marian Thayer, standing with her husband and son, who was told by him that the ship had an hour to live.[8]

In *The Loss of the SS Titanic: Its Story and Its Lessons*, published later that year, Lawrence Beesley, a thirty-five-year-old Cambridge graduate, widower, and former science teacher at Dulwich College in London, described his own jump which took place, he explained, 'under very similar circumstances' to those of Ismay. Lifeboat 13 was in the process of being lowered and the call for women and children had been repeated. 'Just then one of the crew looked up and saw me looking over. "Any more ladies on your deck?" he said. "No," I replied. "Then you had better jump".'

Beesley, a Christian Scientist who was on his way to 'study the greater work in New York', felt no guilt about leaving the ship; the moment of his escape rather served as an epiphany which left him with a visionary sense of purpose. In an article for the *New York Times*, which appeared on 29 April, he defended Ismay's actions and described how 'the moment I realised there was any danger I turned at once to the method and habit which are incumbent on a Christian Scientist – the attempt to eliminate fear from the human mind'. This enabled him, Beesley believed, to 'stand quietly on the deck and watch boats being lowered until the moment came when I was able to get a place . . . without depriving anyone of room'. In another article written in December that year for the *Christian Science Sentinel*, Beesley described how he recalled, as the ship was sinking, 'the spiritual sense of the ninety-first psalm – to "be still, and know that I am with God"'. In his stillness, Beesley expanded to absorb what he called 'the magnitude of the whole thing', soaking it up like a sponge. Not having lost any friends or family and fully covered by an insurance policy he took out at the last minute, Beesley's most dominant emotion was gratitude; watching the *Titanic* go down he was overwhelmed by the 'many things to be thankful for', and he renewed his pact with his maker. He became attuned to being alive; from now on he resolved to live with increased intensity. He also resolved to record the experience. While he was on the *Carpathia*, Beesley wrote a letter to the London *Times* 'urging the taking of immediate steps to insure safety of passengers and pointing out as dispassionately as possible the reasons for the disaster'. Hearing about the document from Officer Lightoller, Ismay asked to see it. 'I hesitated,' Beesley recalled, 'knowing that while it did not seek to affix blame, the deduction would be that there was blame attachable somewhere – where I did not know and I did not wish Mr Ismay to think we were planning to criticise either his officers or his company.' Ismay read and returned the letter without 'raising the slightest objection to it'.

Five weeks after the wreck, while at a lunch party in Boston, Beesley was persuaded by the editor of the *Boston Herald* to try his hand at writing something longer and the publisher Houghlin Mifflin provided him with a room in a residential club in order to get it done. It took Beesley six weeks to produce his masterpiece. A man raised on adventure stories and tales of the sea, Beesley, discarding his journalistic voice, described in novelistic detail the romance of ship life: 'Each morning the sun rose behind us in a sky of circular clouds, stretching round the horizon in long, narrow streaks and rising tier upon tier above the skyline, red and pink and fading from pink to white.' He watches from the top deck 'the swell of the sea extending outwards from the ship in an unbroken circle until it [meets] the skyline with its hint of infinity'.[9] On board, Beesley is as calm and contained as a babe in his mother's arms. Even after the ship has struck the iceberg, he explains that 'to feel her so steady and still was like standing on a large rock in the middle of the ocean'.[10] In order to help his readers to imagine the vertiginous experience of being lowered in a lifeboat, he asked them to measure seventy-five feet of a tall building and then look down.

It is when he describes the *Titanic*'s final moments that Beesley reaches the limits of what can be expressed. 'I realise how totally inadequate language is to convey to some other person who was not there any real impression of what we saw,' and he returns again and again to the inadequacy of language when faced with extremities on this scale. 'No novelist would dare to picture such an array of beautiful climatic conditions,' he says of the sunrise over the icebergs on the morning after the wreck. 'No artist could have conceived such a picture.' His subject, Beesley realises, is no longer the 'lessons' to be learned from the loss of the *Titanic*; he is now writing about inexpressibility. More than this: his subject *has become inexpressible*; to

describe the 'high drama' of what he experienced 'borders', as he puts
it, 'on the impossible'. Beesley's book is not only about the *Titanic*; it
is about *everything*. But his story could also, he says, be contained in
'a single paragraph' and he gives us that paragraph on his opening
page: 'The keel of the *Titanic* was laid on March 31, 1909, and she was
launched on May 31, 1911; she passed her trials before the Board of
Trade officials on March 31, 1912, at Belfast, arrived at Southampton
on April 4, and sailed the following Wednesday, April 10, with 2,208
passengers and crew, on her maiden voyage to New York.'[11]

Beesley wants to produce a work of literature and not a catalogue
of facts. But there is a sense in which the experience of the *Titanic* is
best expressed not in sentences but as a list. It is as a list that the story
first began to shape itself: the passenger list provided the most
popular reading on board, after which the list of survivors became
the most widely read document in the world. Columns analysing the
percentages of deaths in first-, second- and third-class have been
pored over for a century, and we continue to be staggered by the
quantities of employees and objects on board the ship. The story of
the *Titanic* could be written as an inventory. The kitchen and dining-
room staff alone included butchers, bakers, night bakers, Vienna
bakers, the passenger cook, grill cook, fish cook, sauce cook, soup
cooks, the larder cook, the roast cook, Hebrew cook, pastry cook,
vegetable cook, the cook and stewards' messmen, the coffee men, the
assistant confectioner, chefs, the entrée cook, the iceman, the scul-
lions, the plate-warmers, kitchen porters, carvers, scullery men, the
kitchen clerk, wine butler, dining saloon steward, pantry stewards,
plate steward, reception room steward, lounge attendant, smoke
room steward, verandah café steward, à la carte restaurant manager,
maître d', assistant waiters, and the ship's bugler who summoned the
passengers to table. The ship's pantry carried 500 breakfast cups,
3,000 teacups, 1,500 coffee cups, 3,000 beef tea cups, 1,000 cream
jugs, 2,500 breakfast plates, 2,500 dessert plates, 12,000 dinner plates,
4,500 soup plates, 1,200 coffee pots, 1,200 teapots, 4,500 breakfast
saucers, 3,000 tea saucers and 1,500 coffee saucers.

But there is also a sense in which the facts and figures tell us nothing at all, and Beesley duly arranges his version of the loss of the *Titanic* into nine rich and layered narrative chapters and a hundred tightly written pages. Despite his desire to tell a story, Beesley's account pivots and stalls around a handful of words whose floating meanings he cannot pin down. The word 'beauty' particularly bothers him, because beauty has now aligned itself to death. He repeats the word again and again, as though trying to fix it into place: 'the beauty of the night, the beauty of the ship's lines, and the beauty of the lights, all these things in themselves were intensely beautiful'.[12] He also returns to 'motion' and 'motionless', which terms seem to have lost their moorings. 'The *Titanic* lay peacefully on the surface of the sea – motionless, quiet, not even rocking to the roll of the sea . . . the sea was as calm as an inland lake save for the gentle swell which could impart no motion to a ship the size of the *Titanic*.' In Beesley's hands 'motionless' also describes movement. The sea is at the same time swelling and motionless; the ship is motionless when she is both steaming ahead and sinking – 'and there she remained – motionless!' The sky, the floating bodies, the air, the stars, the lifeboats are all motionless – 'the oarsmen lay on their oars, and all in the lifeboat were motionless' – as is the cold, 'if one can imagine "cold" being motionless and still'.[13]

At his desk in the Boston club, Beesley – who has previously written only letters, university essays and school reports – cannot stop writing. He is pouring out page after page of rapt prose: 'The complete absence of haze', he says of the night,

> produced a phenomenon I had never seen before: where the sky met the sea the line was as clear and definite as the edge of a knife, so that the water and the air never merged gradually into each other and blended to a softened rounded horizon, but each element was so exclusively separate that where a star came low down in the sky near the clear-cut edge of the waterline, it still lost none of its brilliance. As the earth revolved and the water edge came up and covered partially the star, as it were, it simply cut the star in two, the upper half continuing to sparkle as long as it was not entirely hidden, and throwing a long beam of light along the sea to us.[14]

Beesley's rhetoric recalls that of the master-mariner Joseph Conrad, and these pages in particular are reminiscent of the first part of *Lord Jim*, in which Jim's pilgrim ship, the *Patna*, moves 'so smoothly that her onward motion was imperceptible to the senses of men, as though she had been a crowded planet speeding through the dark spaces of ether behind the swarm of suns'.[15] At times Beesley's descriptions mirror those of Conrad with such uncanniness as to suggest he was using *Lord Jim* as his model. Beesley writes of being aboard the *Titanic*: 'To stand on the deck many feet above the water lapping idly against her sides, and looking much further off than it really was because of the darkness, gave one a sense of wonderful security.'[16] Conrad writes of Jim on the *Patna*: 'A marvellous stillness pervaded the world, and the stars, together with the serenity of their rays, seemed to shed on earth the assurance of everlasting security.'[17]

Lawrence Beesley, who based his experience of the *Titanic* on adventure stories (and became father-in-law to the novelist Dodie Smith, who wrote *One Hundred and One Dalmatians*) was later turned into a literary figure himself. Julian Barnes describes in *A History of the World in 10½ Chapters* how, when Pinewood Studios filmed *A Night to Remember* on a pond in Elstree Studios, Beesley asked to be an extra in the scene where the ship goes down. Not having the necessary actor's union card, permission was denied. But so determined was he to repeat the pivotal experience of his lifetime that, dressed as an Edwardian, he slipped unnoticed onto the set. 'Right at the last minute,' writes Barnes, 'as the cameras were due to roll, the director spotted that Beesley had managed to insinuate himself to the ship's rail; picking up his megaphone, he instructed the amateur imposter kindly to disembark. And so, for the second time in his life, Lawrence Beesley found himself leaving the *Titanic* just before it went down.'

Contemplation and reflection played no part in Ismay's daily routine. Described by a friend as 'austere, uncompromising and intolerant to the weaknesses of human nature', Ismay resisted the effects of even minor daily disorder and shielded himself from what events he could not control. The loss he now suffered was beyond anything that could be measured. Incapable of articulating, or even comprehending, his response to the wreck, he said nothing on board the *Carpathia*. He detached himself from the experience of Sunday night, he froze himself into the present tense. The disaster, Ismay decided, had nothing to do with him; it happened *around* him, not *to* him and certainly not *because of* him. 'Has it occurred to you,' he was asked at the British inquiry, 'that, except perhaps apart from the Captain, you, as the responsible managing director, deciding the number of boats, owed your life to every other person on that ship?'

'It has not,' Ismay replied, in all honesty. And yet the same friend who had described Ismay's intolerance of human weakness also noted that 'no one could be kinder or more sympathetic or enter more fully into the troubles of others . . . Had I to sum up his character in one word – it would be that of Integrity.'[18]

The only crew member from the *Titanic* to see Ismay on the *Carpathia* was Second Officer Charles Lightoller, the most senior of the four surviving officers. Lightoller, another Christian Scientist, described in an article for the *Christian Science Journal* the following October how 'all fear left me and I . . . realised the truth of being'.

Swept off the ship, Lightoller spent the night, along with twenty-eight other frozen men, straddling a capsized lifeboat. Aged thirty-eight in 1912, he had been at sea since he was thirteen and experienced two shipwrecks before he was twenty-one. An employee of the White Star Line for fourteen years, before joining the crew of the *Titanic* he had been an officer on the *Oceanic*. In a longer account of the disaster, written twenty-three years later, Lightoller describes how, as the ship was sinking, he was standing 'partly' in a lifeboat helping to load the women and children when Captain Smith suggested that, as there were no other seamen available to man the boat, Lightoller should 'go with her'. 'Praises be', Lightoller remembers in horror, 'I had just sufficient sense to say "Not damn likely" and jump back on board.' Ismay's opposite, Lightoller is distinguished for having jumped off a lifeboat and onto a sinking ship. He acted, he said, on 'pure impulse . . . an impulse for which I was to thank my lucky stars a thousand times over in the days to come'.[19]

Lightoller later explained to those who were bewildered by Ismay's behaviour on the *Carpathia*, that the chairman of the White Star Line was 'obsessed' by guilt; as he lay on his bed, a wretched figure, he 'kept repeating that he ought to have gone down with the ship, because he found that women had gone down'.[20] But according to Ismay's account, he did not know that women had gone down on the *Titanic*. Asked at the US inquiry what proportion of women and children were saved, he replied, *I have no idea, I have not asked*.

On the Thursday, 18 April, Dr McGhee suggested that Jack Thayer visit Ismay before the *Carpathia* docked in New York, to 'help relieve the terribly nervous condition he was in'. The seventeen-year-old, who had jumped from the ship and been saved by the same capsized lifeboat as Lightoller, went down to the doctor's cabin immediately and later described how, 'as there was no answer to my knock, I went right in. [Ismay] was seated, in his pyjamas, on his bunk, staring straight ahead, shaking all over like a leaf. My entrance apparently did not dawn on his consciousness. Even when I spoke to him and tried to engage him in conversation, telling him he had a perfect right to take the last boat, he paid absolutely no attention and continued to look ahead with his fixed stare. I am almost certain that on the *Titanic* his hair had been black with slight tinges of grey, but now his hair was virtually snow white.'[21]

Ismay may have been less concerned that he had failed his passengers than that he had failed his father, a legendary man in the shipping world. But what dominated his thoughts was the idea that his ship had failed him. Whatever the size and splendour of your ark, he now realised, the sea is the irreconcilable enemy of ships and men and optimism. This is what Joseph Conrad, who spent half his life at sea, understood: 'When your ship fails you,' he wrote in *Lord Jim*, 'your whole world seems to fail you.' The *Titanic* had been Ismay's pride, his passion, his palace. She was indestructible, the culmination of his father's ambition. She had promised to garland the family name in laurels, to bestow on Ismay honour, glory, and unchallenged dominion of the waves. 'The love that is given to ships,' says Conrad in *The Mirror of the Sea*, 'is profoundly different from the love men feel for every other work of their hands.' It is 'untainted by the pride of possession'. Conrad was describing the love that is given to ships by the crew rather than by the owner. In his memoirs, Lightoller also talked about 'the loyalty between a ship and her crew': 'It is not always a feeling of affection either. A man can hate his ship worse than he can hate a human being . . . Likewise a ship can hate her men.'[22] To the crew, the owner was as distinct from a seaman as a duck from a dolphin; shipowners were not romantic wanderers, they were men of order and control, fastidious pen-pushers ill-suited to the unruliness of the oceans. Men like Lightoller assumed that the owner would see his ship as no more than a description of profit and loss. But Ismay was not typical; he loved the *Titanic* like 'a living thing', as his wife put it. Perhaps, he later wrote to Marian Thayer, he had loved this ship 'too much', had been 'too proud' and such was his punishment. Ismay had grown in stature with his ships. Three years earlier Winston Churchill announced in *The Times* the arrival of 'a new time. Let us realise it. And with that new time strange methods, huge forces, larger combinations – a Titanic world – have sprung up around us.' This was Ismay's world. And now, like a Titan, he had angered the gods.

'All fates', wrote Evelyn Waugh, 'are "worse than death"', and for Ismay the awareness of lost honour was the worst fate of all. There was a difference, he now understood, between surviving and being alive. Lawrence Beesley felt the euphoria of having narrowly escaped death and Lightoller considered his continued life a miracle; but Ismay's attitude suggests that he took for granted that he was living while others were not. He was on the *Carpathia*, he accepted, not because an impulse had caused him to jump, but because his drive to survive had been greater than that of many other men on the *Titanic*. Ismay was profoundly shocked by the mortality of his ship, but unsurprised at his own durability. Comments he later made in correspondence suggest that he was wearied by his continued consciousness as though it were his curse to escape unscathed from every cut-throat situation, his burden to vary from the happy race of men who die when their allotted time is up.

As Ismay lay on his bed in the *Carpathia* on the morning of

Monday 15 April, contemplating as best he could the events of the last twelve hours, the *Olympic* was heading towards them. Built alongside one another in the Harland & Wolff dockyards, the *Olympic* was the *Titanic*'s twin. 'Everything was taken to be doubled,' the chief designer said of their construction. 'Anything which was taken up for one applied to the second ship the same.'[23] The interiors of the two liners were mirror images of one another; passengers on the *Titanic* were provided with maps of her layout which belonged to the *Olympic*. Captain Haddock of the *Olympic* now telegraphed the *Carpathia*: perhaps, he suggested, the *Olympic* might relieve Captain Rostron of his White Star passengers? Rostron's response to Haddock's request was cautious: 'Do you think it advisable the *Titanic*'s passengers see *Olympic*? Personally, I say not.' After a brief exchange with Ismay, he then sent another, more urgent, message to Haddock:

Form No. 4.—100.—17.8.10.   Deld. Date **15 APR 1912**

The Marconi International Marine Communication Co., Ltd.,
WATERGATE HOUSE, YORK BUILDINGS, ADELPHI, LONDON, W.C.

No. **OLYMPIC.**   OFFICE.   **15 APR 1912**   19

Handed in at   **CARPATHIA**   CHARGES TO PAY.

This message has been transmitted subject to the conditions printed on the back hereof, which have been agreed to by the Sender. If the accuracy of this message be doubted, the Receiver, on paying the necessary charges, may have it repeated whenever possible, from Office to Office over the Company's system, and should any error be shown to exist, all charges for such repetition will be refunded. This Form must accompany any enquiry respecting this Telegram.

                                                                    Total

To

    COMMANDER OLYMPIC   RECEIVED 3.22 pm NYT

    MR. ISMAY ORDERS OLYMPIC NOT TO BE SEEN BY CARPATHIA.
    NO TRANSFER TO TAKE PLACE. ROSTRON.

It was 'very undesirable' Ismay later explained to the inquiry, 'that the unfortunate passengers from the *Titanic* should see her sister ship so soon afterwards'.

The liner on which Captain Edward John Smith had died was an age away from the rigged ships in which, aged sixteen, he began his life at sea. A working-class boy from Stoke-on-Trent, Smith's older step-brother, Joseph, was a sea captain who had been captured by pirates; the young Edward longed for similar adventures. To prevent the lad from running away to Liverpool and signing up on the first available barque, Joseph had taken Edward with him on his next voyage. But after years of transporting ossified bird dung and sleeping eight to a cabin with his trunk lashed to the floor by iron rings, Smith fixed his eye on the glamour of the White Star Line. No more oilskins, weevil-infested biscuits and cussing sea dogs; here was romance of a higher order. He was thirty when he joined White Star in 1880, and seven years later he attained the rank of Captain. His first command, the *Republic*, was a hybrid of steam and sail; one lonely smokestack nestled inside a forest of masts. In 1888, Captain Smith was trans-ferred to the *Republic*'s twin sister, the *Baltic*: the White Star Line always built ships in pairs. 'My love of the ocean that took me to sea as a boy has never left me,' he later said. 'In a way, a certain amount of wonder never leaves me . . . There is wild grandeur too, that appeals to me in the sea.'[24]

By 1892, Smith was the most highly respected skipper in the British merchant service and he was rewarded with the honour of taking each new White Star liner on her maiden voyage. Known as the 'millionaire's Captain', he was now a celebrity seaman – travelling under Captain E. J. Smith was as thrilling as being on the *Titanic* itself – and his salary, twice what Rostron was earning with the Cunard Line, reflected his status. Smith was described by Lightoller as 'a great favourite' of any crew, a man everyone wanted to work under: 'Tall, whiskered and broad, at first sight you would think to yourself, "here's a typical Western Ocean Captain. Bluff, hearty and I'll bet he's got a voice like a foghorn". As a matter of fact, he had a pleasant, quiet voice and invariable smile. A voice he hardly raised above a conversational tone – not to say couldn't, in fact.'[25] J. E. Hodder Williams, of the publishers Hodder and Stoughton, who crossed the Atlantic with Captain Smith on many occasions, thought him 'the perfect sea captain'. He 'had an infinite respect for the sea. Absolutely fearless, he had no illusions as to man's power in the face of the infinite.' The American writer Kate Douglas-Wiggin gave an account in her autobiography of how she crossed the Atlantic with Smith over twenty times.

> There were no electric lights then, nor 'Georgian' or 'Louis XIV' suites, no gymnasiums or Turkish baths, no gorgeous dining salons and meals at all hours, but there were, perhaps, a few minor compensations, and I can remember certain voyages when great inventors and scientists, earls and countesses, authors and musicians and statesmen made a 'Captain's table' as notable and distinguished as that of any London or New York dinner. At such times Captain Smith was an admirable host; modest, dignified, appreciative; his own contributions to the conversation showing not only the quality of his information but the high quality of his mind.

Captain Smith offered an uneventful but glamorous voyage; it is part of the strangeness of the sea that a captain who on shore lives a quiet suburban life, becomes an object of fascination when on board his own ship.

On the *Titanic*, the Marconi operators received or intercepted eighteen separate ice warnings; some of these Captain Smith saw, one of them he passed on to Ismay, one he showed to Lightoller, while several never made their way onto the bridge. On the evening of 14 April, while the Captain was at a dinner party given in his honour by a party of American millionaires, the iceberg, whose position he was aware of, was only fifty miles ahead. Captain Smith was losing his fear of the sea: it was the grandeur of the ships rather than that of the water which now gave him cause for wonder. 'When I observe from the bridge', he told a reporter, 'a vessel plunging up and down in the trough of the seas, fighting her way through and over great waves, tumbling, and yet keeping on her keel, and going on and on – I wonder how she does it, how she can keep afloat in such seas, and how she can go on and on safely to port.'[26] When he brought the *Adriatic* to New York on her maiden voyage in 1907, Smith told journalists that 'shipbuilding is such a perfect art nowadays that absolute disaster involving passengers is inconceivable. Whatever happens, there will be time enough before the vessel sinks to save the life of every person on board. I will go a bit further. I will say that I cannot imagine any condition that would cause the vessel to founder. Modern shipbuilding has gone beyond that.'

The most enigmatic member of his crew, it is impossible to account for the final moments of Captain Smith. After informing Ismay that the ship was sinking, he seemed to melt away. Some said that he took up his megaphone and ordered the international population of men and women to 'Be British, my lads', while other witnesses have him variously firing a gun into a crowd, turning the gun on himself, and jumping overboard with a baby swaddled in his arms. In another version, he is last seen in the water swimming alongside a lifeboat; when an oar is held out to him, he says, 'Goodbye boys, I'm going to follow the ship.' There are the inevitable rumours of his survival, including a sighting of him alive and well in Baltimore, but it seems most likely that Captain Smith died on the bridge, where he was standing alone. As he was not wearing a lifejacket, he was at least spared the slow death from hypothermia suffered by most of the others. Mrs Thayer said she saw the Captain after midnight, on the port side of the bridge, and Lightoller, who was working on the port side all night, says that Smith ordered him to start loading women and children into lifeboats from the promenade deck, one level below the boat deck, forgetting that, unlike the *Olympic*, the promenade deck on the *Titanic* was fully glazed with what were known as 'Ismay' screens, a sheltering wall which Ismay suggested at planning stage would protect the promenading passengers from sea-spray. The Ismay screens on the *Titanic* were the only visible difference between the two sisters. The passengers, once they had been herded down to the lower deck, had to be pushed headfirst though the opened windows and into the boats. Captain Smith's confusion about which ship he was on reveals something of his frame of mind.

'It is the great Captain', Smith said, 'who doesn't let things happen', and for forty-three years nothing much had happened on any of his crossings. But latterly, over a period of eight months, he was responsible for two accidents and one near miss. The previous September the *Olympic,* while under his command, had rammed into the British cruiser, HMS *Hawke*; then in February he drove the *Olympic* over a submerged wreck and lost a propeller blade. Then leaving Southampton on 10 April, the *Titanic* narrowly missed crashing into the American liner, *New York*. 'Had he been saved,' the *Washington Times* commented on 17 April, 'Captain Smith's career was over. He had twice escaped the rule that the victim of an accident to a vessel must give up his post,' and his preferential treatment by the White Star Line violated 'a deep sea tradition to dispense with the services of officers in command of vessels that met with disaster'.[27]

Captain Smith was sixty-two when he took command of the *Titanic*. It is often said that this was to be his last voyage before retirement, but he is unlikely to have stepped down at the point in his career where he was at his most assured, and the White Star Line would have been unwilling to let go of such an asset. If anything, Smith's retirement began when the *Olympic* and *Titanic* were launched, at which point he relaxed on the job. 'Either of these vessels could be cut in halves,' he told a young officer, 'and each half would remain afloat indefinitely.'[28] Like his first-class passengers, Smith was pampered, celebrated and over-confident – the very opposite of the stoical Captain Rostron, who at forty-two was beginning his career as a commander. Having trained as a cadet on the Mersey ship, the *Conway*, Rostron joined the Cunard company in 1895 and had been in command of the *Carpathia*, whose plain interior was described in a trade magazine as being 'suggestive of good taste and solid comfort', for three months. Slim, balding and religious, Rostron was praised after the rescue for his 'unaffected valour', his 'own indifference to peril, his promptness and his knightly sympathy'.[29]

As the ship's chaplain held a service of thanksgiving and remembrance, Rostron ensured that the orders he had dispensed earlier that morning were being followed. While the second-class dining room was turned into a hospital for the injured and his own cabin prepared for the most elite of the first-class widows, including Mrs Thayer, the floors and tables of the first- and third-class dining rooms were cleared for further women to bed down and the smoking rooms converted into dormitories for the men. Meanwhile the *Carpathia*'s women formed a relief committee to provide clothing for those who had arrived in their dressing gowns or evening wear, ship's blankets were cut up to make warm coats for the children, and the first-class *Titanic* passengers formed a team of seven to care for those in steerage. Another committee of survivors raised $15,000 to help those who were too destitute to continue with their journeys. Not one of the *Carpathia*'s passengers complained about the inconvenience caused to their cruise. One *Titanic* survivor said that she 'had never seen or felt the benefits of such royal treatment'. In a letter to Walter Lord in 1954, Bertha Watt, who was a child when she was rescued, said that on the *Carpathia* 'I learned a great deal of the fundamentals I have built a happy life on, such as faith, hope, and charity.'[30] Marie Young, a first-class survivor, described how the experience of 'those who mingled freely in the ship's company' was 'richer, by far', than that of Ismay, alone on his bunk. Meanwhile, the *Titanic* crew, whose pay had stopped the minute the ship sank, waited for word from their employer, who had not yet emerged from his private cabin.

At 8 a.m., before being sedated by the *Carpathia*'s doctor, Ismay had scribbled a Marconigram to Philip Franklin, American vice-president of the International Mercantile Marine, the parent company of the White Star Line:

Deeply regret advise you *Titanic* sank this morning after collision with an iceberg, resulting in serious loss of life. Further particulars later.

The message was written on the advice of Rostron, who described Ismay as 'mentally very ill at the time'. So unable was Ismay to function, Rostron said, that 'our purser asked him to add the last three words'. But Ismay did not send further particulars later, and nor was the Marconigram sent that day. Instead, it got buried under the mountain of messages from *Titanic* passengers which the exhausted operator, Harold Cottam, was trying to work his way through.

Philip Franklin, who claimed to know as little as everyone else about the fate of the *Titanic* and her passengers, did not receive Ismay's official notice of the wreck until Wednesday morning, by which point it had been overtaken by a deluge of other messages. During the previous two days, the White Star Line and the press offices were receiving and repeating what the British *Chronicle* described as 'an orgy of falsehood' which began with a Marconigram from an unidentifiable source which Franklin and others received on Monday, informing him that everything was fine:

> All *Titanic* passengers safe. The *Virginian* towing the liner into Halifax.

Delighted, Franklin arranged for a ship to meet the *Titanic* as she came into port, and a train to take her passengers on to New York. Two hours later he received another message from another unidentifiable source, this time saying that it was 'reported' from the *Carpathia* that:

> All passengers of liner *Titanic* safely transferred to this ship and the SS *Parisian*. Sea Calm. *Titanic* being towed by Allan Liner *Virginian* to port.[31]

Confused, Franklin ordered the train making its way to Halifax to turn around and come back again. The *Titanic* passengers would now be met, he assumed, in New York. The *New York Sun* felt able to run the headline 'ALL SAVED FROM THE TITANIC'. It was

only when David Sarnoff, a twenty-one-year-old Marconi operator, picked up a message from the *Olympic* that the truth was
known:

> *Carpathia* reached *Titanic* position at daybreak. Found boats and
> wreckage only. *Titanic* had foundered about 2.20 a.m. in 41.16 north,
> 50.14 west. All her boats accounted for. About 675 souls saved, crew
> and passengers included.

Sarnoff passed the information on to the Press Association, after
which, in his own words, 'Bedlam was let loose'. A second
Marconigram from Captain Haddock of the *Olympic* read:

Yamsi was Ismay's codename. Until that moment, Franklin said, 'we
considered the ship unsinkable, and it never entered our minds that
there had been anything like a serious loss of life'. The *New York Times*
then ran the headline: 'MANAGER OF THE LINE INSISTED SHE
WAS UNSINKABLE EVEN AFTER SHE HAD GONE DOWN'.

By Tuesday the names of survivors had begun to slowly trickle
into the news offices, sent via the *Olympic* because the *Carpathia* had

only a short-range wireless. Anguished crowds waited for news
outside the White Star Offices in New York, Liverpool, Southampton
and London, two wives of the crew dying from the shock and
suspense. Word spread that Ismay was alive while other prominent
men were not and that the *Titanic* had received, and ignored, several
ice warnings. It was believed that the source of the reassuring
Marconigrams received by Franklin, and also by Congressman
J. A. Hughes – whose newly married daughter, Mrs Lucian P. Smith,
had accused Ismay of demanding a private cabin on the *Carpathia* –
was Ismay himself.

The papers on the morning of Tuesday 16 April offered opinions on
the wreck by experienced seamen. Admiral Dewey was quoted in the
Washington Post as saying that any passenger who crossed the North
Atlantic in a transatlantic vessel 'takes his life in his hands . . . the greed
for money-making is so great that it is with the sincerest regret that I
observe that human lives are never taken into consideration'. Admiral
F. E. Chadwick wrote in the *New York Evening Post* that the '*Titanic*
was lost by unwise navigation, by running at full speed'. The *New York
Times* revealed what everyone in the shipping world already knew: the
Titanic had been allowed, by the British Board of Trade, to go to sea
with insufficient lifeboats. 'If the *Titanic* had been under United States
Government supervisions,' a naval contractor wrote in the San
Francisco *Examiner*, 'its owners would have been compelled to equip
it with forty-two lifeboats at least . . . The trouble with the English
regulations is that they are behind the times.' The 'old fogeyism' of the
British had cost American lives, and 'we Americans', as Admiral Dewey
put it, 'surely have some rights in this matter'. The US naval historian,
Rear Admiral Alfred Thayer Mahan, a relation of Marian Thayer's dead
husband, wrote in the *Evening Post:* 'I hold that under the conditions,
so long as there was a soul that could be saved, the obligation lay upon
Mr Ismay that that one person and not he should have been on the
boat.' An editorial in New York's *Truth* commented with irony: 'I
cannot help regarding it as "providential" that the chairman of the
company happened to be standing where he was at the moment when
the last boat – or was it the last but one? – left the ship, and there were

no women or children at hand to claim the place into which he was
thus enabled to jump.'

In the Senate, William Alden Smith of Michigan called for a
formal inquiry to be conducted with immediate effect. His resolution was given unanimous support; he was to authorise a panel
composed of an equal number of Republicans and Democrats 'to
investigate the causes leading to the wreck of the White Star liner
Titanic, with its attendant loss of life so shocking to the outside
world'. Senator Smith was empowered to 'summon witnesses, send
for persons and papers, to administer oaths, and to take such testimony as may be necessary to determine the responsibility therefore'.
The sinking of the *Titanic*, said Senator Rayner of Maryland, was a
crime and should be investigated in the same way as any other
crime. Had the *Titanic* been an American ship and 'subject to our
own criminal procedure', Ismay would be convicted of 'manslaugh-
ter if not murder'. 'Here you have the spectacle of the head of a line
failing to see that his ship is properly equipped with life-saving
apparatus, heedless of the warnings that he was sailing in dangerous
seas, forsaking his vessel, and permitting 1,500 of her passengers and
crew to be swallowed by the sea. Mr Ismay,' Rayner suggested to the
Senate, 'the officer primarily responsible for the whole disaster' has

reached his destination in safety and unharmed. Mr Ismay should be
brought here and be made to explain these things. He should not
be requested to come. It is not a question of his good will . . . and he
should be asked particularly to explain how he, the directing manager
of the company, the superior of the Captain, and not under the
Captain's orders, directed the northern route which ended so fatally and
then left hundreds of passengers to die while he took not the last boat,
but the very first boat that left the sinking ship . . . All civilised nations
will applaud the criminal prosecution of the management of this line. If
they can be made to suffer, no sympathy will go out for them.

Ismay, meanwhile, was planning his return voyage. He could not face New York: the onslaught of information, the questions, the crowds, the grief, his wife's American family, the White Star Line executives, the insurers, the press, the publicity, the sharpening sense of the horror of it all. He would have to endure, he assumed, a rough few days – possibly even weeks – of media coverage before interest in the *Titanic* would fade away and he could get his life back.

Escape came in the form of another White Star liner, the *Cedric,* currently docked in New York and ready to sail. On disembarking from the *Carpathia*, Ismay and the 200 surviving *Titanic* crew members could immediately board the *Cedric* and be home within the week. As fog was delaying the *Carpathia*, the need to postpone the *Cedric*'s imminent departure quickly became a fixation for Ismay. It would not be absconding to return to England straight away, he reasoned; it was simply business as usual: he was, quite rightly, picking up the rhythm of his life again and the crew would need to get themselves positions on new ships. So on Wednesday morning at 9 o'clock Ismay sent a Marconigram to Franklin reading:

> Very important you should hold *Cedric* daylight Friday for *Titanic* crew. Answer, Yamsi.

Messages addressed to 'Yamsi' would be delivered personally to Ismay, and messages signed 'Yamsi' indicated that they were sent personally by Ismay himself. The name was not, Franklin later explained, 'used by us very much over here' – he had never himself used the name before – while it was 'used entirely on the other side'.[32] But when Ismay called himself Yamsi on the *Carpathia*, where he was on neither one side nor the other, it took on a different significance. Ismay backwards, Yamsi was an inversion of the man Ismay used to be. He had stepped into a looking-glass world where, as Lewis Carroll's Alice explains when she 'softly jumps' from the fireplace into the drawing room on the other side of the mirror, everything is the same 'only the things go the other way'.

The official Marconigram that Ismay had been persuaded to send Franklin on Monday morning was eventually received at 9 o'clock on Wednesday morning. No sooner had Franklin absorbed its contents than 'Yamsi's' message arrived. So rather than being furnished with the 'further particulars' promised in the first message, Franklin was instead being told to delay the *Cedric*. He replied:

> Accept my deepest sympathy horrible catastrophe. Will meet you aboard *Carpathia* after docking. Is Widener aboard?

George Widener, who had not survived, was the son of a director of the International Mercantile Marine. Franklin also forwarded a message from Ismay's wife, Florence:

> So thankful you are saved, but grieving with you over the terrible calamity. Shall sail Saturday to return with you.

Ignoring both messages, Ismay repeated to Franklin:

> Most desirable *Titanic* crew aboard *Carpathia* should be sent home earliest moment possible. Suggest you hold *Cedric*, sailing her daylight Friday unless you see any reason contrary. Propose returning in her myself. Please send outfit of clothes, including shoes, for me to *Cedric*. Have nothing of my own. Please reply, Yamsi.

Franklin replied that 'we all consider it most unwise to delay *Cedric* considering all circumstances', adding that he had arranged for the crew to return home on Saturday on the *Lapland*. He later explained that 'we determined it would be a very unfortunate thing to attempt to hold the *Cedric* and hurry the crew on board or agree to Mr. Ismay's sailing under the present circumstances, with which Mr Ismay, as we knew, was not in any way familiar. We were here, and we were hearing the criticism. We knew what was being said, but Mr Ismay had no knowledge or information regarding that. We realised the necessity of getting the crew off, which was just what we wanted done in every other case of the kind and what every shipowner would do.'[33]

Ismay then sent two further Marconigrams which Franklin received one after another between 8.00 and 8.44 a.m. on Thursday 18 April:

> Think most unwise keep *Titanic* crew until Saturday. Strongly urge detain *Cedric* sailing her midnight, if desirable. Yamsi

> Unless you have a good and sufficient reason for not holding *Cedric*, please arrange to do so. Most undesirable have crew New York so long. Yamsi.

Only in his next message, also sent that day, did Ismay mention his wife and answer the question Franklin had asked several Marconigrams earlier:

> Widener not aboard. Hope to see you quarantine. Please cable wife am returning *Cedric*. Yamsi.

Franklin's frantic response, despatched at 4.45 that afternoon, read:

> Concise Marconigram of actual accident greatly needed for enlightenment public and ourselves. This most important. Franklin.

Neither he nor the public would be enlightened until the *Carpathia* reached New York at 9 o'clock that night. It was Captain Rostron's decision to ensure that messages to and from the families of *Titanic* survivors were given priority, to allow no account of the disaster to be transmitted by Marconigram and to ignore requests from the press for further details. It was assumed on shore that the embargo on information had been enforced by Ismay.

The US Navy, picking up 'Yamsi's' messages to Franklin, forwarded them to Senator William Alden Smith in Washington. Glancing over their contents, the Senator made an appointment to see President Taft to arrange that Ismay and other key witnesses be prevented from absconding, and Taft ordered that a Treasury Revenue cutter intercept the *Carpathia* before she docked. It seemed wisest, Senator Smith argued, to begin the inquiry into the wreck of the *Titanic* in New York where the witnesses could all for the moment be found, and then remove it to the capital where the witnesses could be under the jurisdiction of the Senate. In reply to the question of whether he was going to arrest J. Bruce Ismay, the Senator told journalists that 'We not going into this matter with a club. We will proceed cautiously and conservatively.'[34] On Thursday evening, William Alden Smith and his colleague, Senator Newlands, waited at the Cunard Pier for the *Carpathia* to arrive.

The days spent on the *Carpathia* were suspended in a time of their own. In a private account of the wreck written for his scrapbook, the tennis champion Karl Behr, who had lost none of his party, wrote that 'although the sinking of the *Titanic* was dreadful, to my mind the four days among the sufferers on the *Carpathia* was much worse and more difficult to try and forget'.[35] While babies howled for their absent mothers, mothers wept for their dead children and wives grieved for their husbands, Marian Thayer was reunited with her son, Jack, whom she had imagined drowned, and the two socialites, Edith Russell and Lady Duff Gordon, who had lost nothing other than several trunks of couture, joked that 'pannier skirts and Robespierre collars were at a discount in midocean'. Friendships formed, perspectives on the fatal night's events were pulled together, and rumours commenced. The main topic of conversation between the women in first-class was the survival of Bruce Ismay, currently being cared for by the doctor in his private cabin.

On the morning of Wednesday 17 April, while Ismay was writing his Yamsi telegrams, a first-class passenger, Emily Ryerson, let it be known to her friends that at 5 o'clock on Sunday afternoon she had been persuaded by Marian Thayer to take some air. Mrs Thayer had been returning, with her husband and her eldest son, from Berlin where they had been staying as guests of the American Consul General. Mrs Ryerson, who had been on a shopping trip in Paris, was returning for the funeral of her eldest son, a student at Yale who had been killed the previous week in a motoring accident. She had spent the last four days on the *Titanic* in her stateroom. After walking for an hour on the covered deck, Mrs Thayer and Mrs Ryerson stopped in the companionway to watch the sun go down while their husbands took a stroll. The evening was clear, windless and deadly cold; the sky was 'quite pink'. The intimacy between the two women was interrupted by the arrival of an impeccably dressed Ismay, who asked if their staterooms were comfortable. Ismay, with whom Emily Ryerson had only the slightest acquaintance – 'he was a friend of a number of friends of mine' – then 'thrust a Marconigram at me saying we were in among the icebergs. Something was said about speed and he said that the ship had not been going fast but that they were to start up extra boilers that afternoon or evening.' Mrs Ryerson saw that the message also mentioned a distressed liner, the *Deutschland*, which had run out of coal and needed towing. Ismay, Mrs Ryerson said, scoffed at the suggestion that the *Titanic* might come to the rescue, declaring that 'they had no time for such matters as our ship wanted to do the best and something was said about getting in on Tuesday night' (the Marconigram had not in fact been a request for help; its purpose was only to inform the *Titanic* of the *Deutschland*'s position). Mrs Ryerson carried on the conversation 'to keep the ball going'. She was bored by Ismay's company and only half-listening to what was being said, later recalling the 'impression' rather than the 'exact words'. Ismay, she thought, had been speaking 'partly' to Mrs Thayer but 'mostly' to her. When asked about the incident, Ismay – who was possessed, several of his employees noted, of a remarkable memory – remembered only the presence of Mrs Thayer.

When Mr Ryerson and Mr Thayer reappeared, Ismay departed.

His manner throughout, Emily Ryerson concluded, 'was that of one of authority and the owner of the ship and what he said was law'.[36] But it seems more likely that this was an example of Ismay, in his awkward way, being light-hearted. Never good at small talk, or talk of any kind, he was drawn to the openness of certain American women, and everyone was drawn to the sympathetic Marian Thayer. The Marconigram he showed them was the one from the *Baltic* which he had been carrying in his pocket since lunchtime, when the Captain had passed it to him. The message, which would become the focus of the British inquiry, read:

> Have moderate variable winds and clear fine weather since leaving. Greek steamer *Athenai* reports passing icebergs and large quantity of field ice today in latitude 41.51 north, longitude 49.52 west. Last night we spoke German oil tank *Deutschland*, Stettin to Philadelphia not under control short of coal latitude 40.42 north, longitude 55.11, wishes to be reported to New York and other steamers. Wish you and *Titanic* all success.

The suggestion that Captain Smith and Ismay had known about the ship's proximity to ice was shocking. 'I remember with deep feeling,' wrote Lawrence Beesley,

> the effect this information had on us when it first become generally known on the *Carpathia*. Rumours of it went round on the Wednesday morning, grew to definite statements in the afternoon, and were confirmed when one of the *Titanic* officers admitted the truth of it in reply to a direct question . . . It was not then the unavoidable accident we had hitherto supposed: the sudden plunging into a region crowded with icebergs which no seaman, however skilled a navigator he might be, could have avoided . . . It is no exaggeration to say that men who went through all the experiences of the collision and the rescue and the subsequent scenes on the quay at New York with hardly a tremor, were quite overcome by this knowledge and turned away, unable to speak.[37]

A committee of twenty-five surviving passengers prepared a statement to be issued to the press once they reached New York. 'In
addition to the insufficiency of lifeboats, rafts &c, there was a lack of
trained seamen to hand the same – stokers, stewards &c are not efficient boat handlers. There were not enough officers to carry out the
emergency orders on the bridge and to superintend the launching
and control of the lifeboats, and there was an absence of searchlights . . . We suggest that an international conference should be
called [and] we urge the United States Government to take the initia-
tive as soon as possible.'[38]

The *Carpathia,* wrote Lawrence Beesley, 'returned to New York in
almost every kind of climatic condition: icebergs, ice fields and
bitter cold to commence with; brilliant warm sun, thunder and
lightning in the middle of one night (and so closely did the peal
follow the flash that women in the saloon leaped up in alarm saying
rockets were being sent up again); cold winds most of the time;
fogs every morning and during a good part of one day, with the
foghorn blowing constantly; rain, choppy sea with the spray blowing overboard and coming in through the saloon windows'.[39] It was
in drizzle that she steamed past the Statue of Liberty on the
evening of Thursday 18 April. The newspaper boats clustered
around the ship, shouting up questions through their megaphones.
Through his own loudspeaker, Rostron announced that anyone
trying to come on board would be 'shot down'. At New York
harbour's Pier 54, a silent crowd of 30,000 waited as the ship crept
up the river. The only lights visible were the *Carpathia*'s portholes
and the bursts coming from photographer's lamps. The business of
docking, always slow, seemed interminable as the tugs worked away
to get the ship warped in. At 9.30 p.m., the *Titanic*'s survivors,
seventy of whom were widows, began to descend and the crowd
divided into two long, cordoned-off lines through which they
could pass. 'Every figure, every face seemed remarkable,' a

journalist wrote. Senator Smith described the appearance of 'the almost lifeless survivors in their garments of woe – joy and sorrow so intermingled that it was difficult to discern light from shadow'.[40] First to appear on the gangway were the richest passengers; last off were those in steerage including the Lebanese immigrants who had shared Ismay's boat. Finally, six orphaned babies were carried out in the arms of the *Carpathia* crew. The *Titanic* survivors, many of whom were finding speech difficult, were not expecting such a reception. It was the tolling of bells and the booming of cannon which brought home that they 'had passed through a history-making disaster'.[41] Ambulances were waiting for the injured, and the Women's Relief Committee were ready to distribute clothes and shoes among the steerage passengers. Boarding houses were thrown open for those hundreds who had nowhere to stay, while the White Star Line had arranged for passengers who were now destitute to reach their final destination.

When everyone else had left the ship, Philip Franklin, Senator Smith and Senator Newlands showed their passes and slipped quietly on board. The two senators waited impatiently outside Dr McGhee's cabin while Franklin went in to see Ismay with a new suit of clothes, the pair of shoes he had asked for and a fashionable scotch cap. Ismay dressed himself while Franklin explained that he would not be returning home on the *Cedric* and took him through the draft of an official statement he had prepared for Ismay to give to the press. After some minutes, Senators Smith and Newlands demanded to see Ismay, and Franklin replied that he was too ill to be interviewed. 'I'm sorry,' said Senator Smith, 'but I will have to see that for myself', and pushing open the door with his umbrella, informed Ismay that he would be appearing the next morning at the US Senate official investigation into the wreck of the *Titanic*. William Alden Smith had secured his star witness.

Flanked by detectives, Ismay left the *Carpathia* at 11.15 p.m. and went to the rear of the dock to the Cunard offices where an assortment of selected pressmen waited to finally get their story. Franklin's statement was read out by one of the White Star Line officers:

In the presence and under the shadow of a catastrophe so overwhelming, my feelings are too deep for expression in words. I have only to say that the White Star Line, its officers, and employees, will do everything possible to alleviate the suffering and sorrows of the survivors and the relatives and friends of those who have perished. The *Titanic* was the last word in shipbuilding. Every regulation prescribed by the British Board of Trade had been complied with. The master, officers, and crew were the most experienced and skilled in the British service. I am informed that a committee of the United States Senate has been appointed to investigate the circumstances of the accident. I heartily welcome the most complete and exhaustive inquiry, and any aid that I or my associates or our builders or navigators can render is at the service of the public and the governments of both the United States and Great Britain. Under these circumstances, I must respectfully defer making a further statement at this time.

Boasts about the *Titanic*'s superiority as a ship were of no interest to the press, who had already decided that their story was to be a stirring narrative of chivalry and cowardice. 'On what boat did you leave the *Titanic*?' one journalist asked. 'What do you mean?' Ismay replied, unaware that this was an issue. 'I don't know what you mean. I left on a boat leaving from the centre.' What was the number of the boat on which you left, in their order of departure? 'I left from the starboard forward collapsible, the last boat to leave.' Do you want to answer the charge that it is the custom, on the maiden voyage of a new liner, to make as fast a passage as possible in order to secure the good advertising which would follow? 'That statement is absolutely false,' Ismay replied 'with more animation', the *New York Times* reporter noted, 'than he showed at any time during the interview'. 'I can speak for the White Star Line that such a proceeding is not the case, and that the *Titanic* at no time during her voyage had been at full speed.' He was asked how long it took the ship to sink ('two hours and twenty-five minutes since the collision'), whether it was 'true that she remained afloat long enough to save all had there been enough boats' ('I decline to answer'). He explained that he had been asleep at the time of the accident, that he had then come on deck, that he did all he could; he answered questions about the bulkheads and the length of time the lights remained on after the collision. When asked how he happened to be one of the 'mostly women and children' in the lifeboats, he spoke only of the magnificent behaviour of the crew. He said he did not see the ship go down, that he could offer no suggestions as to why the vital wireless message he sent to Franklin from the *Carpathia* had been delayed. When he was asked again how he came to be among the survivors, Franklin intervened to say that the question was unfair. The interview came to an end, and, flanked by bodyguards, Ismay was driven to the Ritz Carlton Hotel for the night.

Amongst those in New York who had booked their passage to Southampton on the *Titanic*'s return journey was the English writer John Galsworthy, who had been rehearsing his new play, *The Pigeon*. Caught up in the moment, Galsworthy decided not to go home immediately, but to attend the *Titanic* inquiry instead.

Youth

This could have occurred nowhere but in England, where men and
sea interpenetrate, so to speak, the sea entering into the life of most
men, and the men knowing something or everything about the sea,
in the way of amusement, of travel, or of bread-winning.

Joseph Conrad, *Youth*

Ismay's father, Thomas Henry, was born in 1837, the year Queen
Victoria came to the throne and Bruce, like the future Edward VII,
was as different from his mighty parent as it was possible to be.
Thomas was a Victorian, Bruce an Edwardian; the father stood for
entrepreneurial strength and imperial greatness, the son for decline.

The Ismays were of Cumbrian stock and Thomas began his life in
the small town of Maryport at the mouth of the River Ellen, in one
of a row of tiny cottages called Whillan's Yard. His own father,
Joseph, was a shipbuilder and after the birth of three further children
– Charlotte and Mary, who were twins, and Sarah – the family moved
to a larger house in the shipyard where Thomas's grandfather, Henry
Ismay, who had once been a sea captain, now worked as a timber
merchant. In their new home, called The Ropery because the ropes
for the ships lay all around, a final child, John Sealby, was born.
Thomas spent his childhood at the harbour, chatting to the sailors
and carving model ships from driftwood. Because he was bright and
promising, he was sent to a good school in Carlisle where he excelled

at sport, especially cricket, and took special studies in navigation. He also improved in writing: a letter from his father, dated 1849, congratulates Thomas on his most recent correspondence which had been 'very well Wrote and spelled'.

Known as 'Baccy' for the tobacco he chewed all day in imitation of the sailors at the harbour, Thomas was short, confident, convivial and popular. He was, a contemporary remembered, 'a dark complexioned lad with dark piercing eyes, whose hobby was the sea, whose ambition was a sea-faring life, and who never seemed so happy as when engaged in fashioning a miniature sailing vessel with a pocket knife out of a block of wood, rigging it with masts and sails . . . and then sailing [it] on the pond at Irthington'. When he was thirteen his father died and Thomas became head of the family. He stayed on at school until he was fifteen and at sixteen began an apprenticeship in the Liverpool shipping firm of Imrie, Tomlinson, where he befriended William Imrie, a fellow apprentice and the son of the joint owner.

At eighteen, he left England to test the fibre of his stuff aboard the *Charles Jackson*, a small sailing ship of 352 tons destined for Chile. This was Thomas Ismay's first sea voyage and, always a keen travel diarist, his record of the journey proves him a solid, humorous and uncomplicated team player.

> January 7, Monday. This is the 19th anniversary of my birthday, and a beautiful day it is, being almost calm, remained on deck nearly all day shooting gulls. The crew in the forecastle had a bottle of brandy given them to drink, and if I judge from the songs I heard them singing, they enjoyed the contents. During the evening Rapp, the Captain and myself were amusing ourselves with singing, Home Sweet Home, etc. Of course, the performance would not have elicited great applause from an audience endowed with taste. I do hope I may enjoy every anniversary as well.

He relieved the Captain's toothache ('I induced him to get a little salt heated, and put into a flannel stocking and tied round the jaw'), guffawed when he got a wetting ('everyone generally enjoys a laugh when another gets wet, and it is best for the sufferer to laugh too') and made the most of the worst conditions ('You have to hold on to your plate to keep it near you, to hold onto your glass of water to avoid the unnecessary luxury of a shower bath'). Once the *Charles Jackson* arrived in Chile, Thomas visited the theatre, enjoyed the food and drink, rode out in the mornings and danced by night with the local girls, who may have 'spat' and 'smoked' but 'between the ankles and the chin', he thought, were 'the best formed race of women in the world'.[1]

When he returned to Liverpool in 1858, Thomas was unstoppable. Aged twenty-two, he set up a ship-broking business, became director of his own line of steamships carrying cargo and passengers to South America and married Margaret Bruce, daughter of local shipowner Luke Bruce. Aged thirty, he bought for £1,000 the goodwill and house flag of the then bankrupt White Star Line of Australian clippers, founded twenty years before by the Liverpool businessmen Henry Threlfell Wilson and John Pilkington, then cashing in on the gold rush. Australian gold was small fry: Thomas Ismay's aim was to cash in on the floods of immigrants making their way to the United States.

Work was his passion and his calling. Ismay Senior wanted to make money but also to make his mark, to bestow pride on his nation and to sire a dynasty who would carry his torch. In 1868 he attended a dinner party at Broughton Hall, the Liverpool home of a Hamburg Jew called Gustavus C. Schwabe. The guests included Thomas's former fellow apprentice, William Imrie. Over a game of billiards, so the story goes, it was agreed that Ismay and Imrie would together form a new company, Ismay, Imrie & Co., which would run steam-powered ships across the North Atlantic and that Harland & Wolff, co-owned by Schwabe's nephew in Belfast, would be the exclusive builders of these vessels. Harland & Wolff would use only the best materials available, and payment would be an unusual arrangement based on cost plus agreed profit margin. Schwabe, who had already invested in another Liverpool shipping company, the Bibby Line, three of whose ships had been built by Harland & Wolff, would support this new venture and ensure further backing from other Liverpool businessmen.

Thus the Oceanic Steam Navigation Company, better known as the White Star Line, was formed, with Ismay, Imrie & Co. as the parent company. Ismay would look after the steam vessels and Imrie the sailing ships, both of which would be built by Harland & Wolff. 'The story of the association between the Belfast builders and the White Star Line,' read an article in the journal *Engineering* in 1912, 'practically involves the story of the development of the Atlantic liner.' It was to be 'a wonderful story', reflected the *London Opinion* in 1904, 'of tonnage, horse-power and names ending in -ic'.

Edward Harland had come to Belfast from Yorkshire in 1854 to take charge of a shipyard on Queen's Island. In 1858 he bought the yard and in 1861 he and Gustav Wolff formed Harland & Wolff. Twenty years later what had begun as a tiny works on a few acres, with one berth and forty-eight employees, contained six slips and a thousand workers including interior designers, artists, tapestry makers and upholsterers. Eventually Harland & Wolff would cover 80 acres, employ 16,000 local men and distribute £28,000 in weekly wages.

———

In 1836, a scientific writer called Dionysius Lardner gave a lecture on steam navigation in which he suggested that 'establishing a steam intercourse with the United States' was as likely as 'making a voyage from New York or Liverpool to the moon'. That same year, Isambard Kingdom Brunel formed the Great Western Steamship Company in order to build a line of steamships to travel between Bristol and New York. Two years later the *Sirius,* owned by the St George Steam Packet Company, was the first ship to cross the Atlantic by steam power alone. Taking eighteen days, she beat Brunel's *Great Western* by twenty-four hours. When the fuel on the *Sirius* ran out, the crew burned the furniture; the science fiction writer, Jules Verne, was inspired to use the scene in *Around the World in Eighty Days*.

Then in 1840 a Canadian named Samuel Cunard beat both the Great Western Steamship Company and the St George Steam Packet Company in winning a British government contract to run the first regular mail and passenger transatlantic steamship service. Cunard's fleet of four sister steamers could offer what no other shipping line was able to do: a swift and punctual schedule of departure. This was a hugely attractive feature for passengers used to waiting at the port until the ship was filled and the winds were fair, before embarking on a passage of indeterminate length. Under sail it took around forty days to reach America; Cunard could reduce this to a fortnight.

The *Britannia,* the first Cunarder to be launched, was celebrated as the last word in shipbuilding and in 1842 Charles Dickens and his wife booked their passage to Boston on board the new liner. The 'stateroom', Dickens recorded, was not at all the 'room of state' pictured in the brochure, with its almost 'interminable perspective'. It was instead 'an utterly impracticable, thoroughly hopeless, and profoundly preposterous box' into and out of which he and his wife had to twine themselves 'like serpents'. Their luggage could no more be 'got in at the door . . . than a giraffe could be persuaded or forced into a flower-pot'. The sumptuous bed they had seen in the illustration in the booking office was a 'very thin mattress' spread over 'a most inaccessible shelf' like a 'surgical plaster'. Reading in bed was not possible – at least, Dickens found, it was not possible to tell what it was you were reading – and nor was it possible to get away from the 'extraordinary compound of strange smells' which accompanied sea life.

Steam, it was believed, would never successfully replace sail. Sail might be slow but it was at least a proven means of transport; by contrast the behaviour of these new-fangled steam wagons was terrifyingly unpredictable and uncertain. Should they break down in mid-ocean, where were the engineers to do the repairs? The fact that the first steamships still carried sails suggested to passengers that the shipbuilders themselves lacked trust in these vessels. Harland &

Wolff were to change all this: their reputation for efficiency and expertise meant that the shift from sail to steam, from wood to iron, and eventually from iron to steel, was made possible. 'All through this great shipyard,' wrote Bram Stoker in an article on Harland & Wolff, 'the biggest and finest and best established in the world, there is omnipresent evidence of genius and forethought; of experience and skill; of organisation complete and triumphant.' The 'perfection of the business organisation' was made apparent to Stoker when he watched the workers clock off at 5.30 on a Friday afternoon and queue to collect their wages. All 16,000 men were paid 'within ten minutes'.[2]

The first liner built for the White Star Line by Harland & Wolff was the *Oceanic*, followed, in under two and a half years, by the *Atlantic, Baltic, Republic, Adriatic, Celtic, Gallic* and *Belgic*. When the *Oceanic* was launched, in 1870, she made news by being the first ship to exceed in length Brunel's *Great Eastern*, which had laid the transatlantic cable and which, at 673 feet, was then the biggest moving object in the world. In *A Floating City*, a novel published in 1867, Jules Verne described crossing the Atlantic on the *Great Eastern*. His narrator is less interested in seeing America than in seeing the ship itself which, he says, was 'something more than a ship, it was . . . a section detached from English soil which, having crossed the sea, united itself to the American continent'. The *Great Eastern*, wrote Verne, was a 'microcosm' carrying 'a little world along with it', containing 'all the instincts, follies, and passions of the human race'.[3]

The *Oceanic*, with berths for 2,000 people, carried only sixty-four passengers on her maiden voyage from Liverpool to New York in March 1871, but it was widely agreed that she was the finest vessel on the Atlantic route. Sleek and elegant, she looked more like a yacht than a steamer. The *Oceanic* was the first modern ocean liner: Thomas Ismay had set a new standard in shipbuilding which every steamship company then followed. It took seven days for her to reach New York, but speed was less important to Ismay Senior than splendour and comfort; his achievement was to build steady ships and thus bring to an end the misery of sea travel. Not only did White Star passengers have no reason to dread their crossing; their journey was now so restful that they were in no hurry to arrive. The designers at Harland & Wolff devised a plan not previously tried before: they placed the first-class staterooms amidships where there was least motion, rather than near the stern where the ship's heaving and vibration were at their greatest. This way they prevented the condition described by Dickens of 'not ill, but going to be', and the compound of strange smells from the galley of which Dickens also complained could additionally be avoided. Staterooms were now double the size and with larger portholes; before fitting the ship out, William Pirrie, a young draftsman of genius employed at the shipyard, was sent on a tour of English and Continental hotels in order to observe their finest features. Where Samuel Cunard was dour and efficient, Thomas Ismay was decadent and efficient; and when the *Baltic* created a new speed record and the White Star won a rival contract to transport the mails to America, the Cunard Company was its only competitor.

Brunel's *Great Eastern* might be a city, but White Star liners were utopias. In *Travelling Palaces,* published in 1913 – seventy years after Dickens travelled on the *Britannia* and one year after the *Titanic* went down – J. A. Fletcher described the new luxury liners as idealised communities whose citizens shared a common goal. While the average home was still without adequate lighting or convenient sanitation, luxury liners provided both in excess. Policemen and courts of law were unnecessary to this new society because 'the state, as epitomised by the liner, takes charge of the passenger'.[4]

Under Thomas Ismay's government, those in first-class would want for nothing, every need was anticipated, every desire satisfied. There were no crowds, no queues. The romance of the sea associated with sails and wooden hulls might be a thing of the past, but Ismay Senior replaced it with a fantasy of a different sort: the sea became an occasion for a party; his voyages provided passengers with unlimited pleasure, abundance and excess. Life on board was a series of concerts, dinners, dances and fancy-dress balls. The dining room now stretched from one side of the ship to the other, allowing passengers enough space to select their own company; they ate not on rows of bolted-down benches but at separate tables provided with *à la carte* menus and elegant, free-standing chairs. Couples could, for the first time in transatlantic history, dine *à deux* in a room divided by partitions which removed them from the site of the central table where the Captain entertained his guests and surveyed his clientele. They might then walk under the shelter of a promenade deck rather than be buffeted about by wind and spray. Love could blossom, business associations form. Oil lamps soon replaced gas and electric bells were installed in every room so that the steward, instead of being shouted for, could be signalled at the press of a button; rows of imitation windows provided the illusion of natural light.

Most importantly, the vastness of the White Star liners made it possible to feel alone and passengers were able to enjoy the contemplation which comes from travel. With coal fires in the grates and curtains on the glass windows, the self-contained staterooms allowed families to imagine they were in their own apartments. Should they prefer not to join in the entertainments, they could cross the Atlantic in complete isolation.

Disaster struck in 1873, when the White Star steamship *Atlantic* ran out of coal on her way to New York and, changing course to refuel in Halifax, hit a rock. Most of the lifeboats were swept away and 250 lives – a third of the passengers – were lost. It was the worst peacetime shipwreck to be recorded at sea, and the press had a field day when the subsequent inquiry found the White Star Line negligent. It was a trauma from which Thomas Ismay would never fully recover and were it not for his determination to forge ahead – he always maintained that the ship's coal supply had been sufficient – the company might have been destroyed by the verdict.

The force of Ismay Senior's personality eclipsed that of his partner, William Imrie, a quiet, modest and cultured man known as 'the prince of shipowners'. Thomas Ismay became famous in the shipping industry for having never offended a rival, for creating thousands of jobs in Belfast, for offering the use of his entire fleet to Queen Victoria for service during the Boer War, for founding a training ship, the *Indefatigable*, which began the shipping careers of over 2,000 boys from poor homes, and for starting a pension fund for Liverpool sailors. He was chairman of the Board of Trade Life-Saving Appliances Committee, Justice of the Peace for Lancashire and Cheshire, a member of Lord Hartington's Commission on the organisation of the Army and Navy and of the Royal Commission on Labour, and he was Deputy Lord Lieutenant and High Sheriff of Cheshire. He turned down as many positions and titles as he accumulated, whether the chairmanship of the London and North-Western Railway Company, the suggestion that he stand for Parliament (he was a Liberal Unionist) or, in 1897, the offer of a baronetcy. In a letter to Bruce, Thomas compared himself to a 'Mr Whiby in the House who declined all honours, preferring to remain Mr Whiby which he said was good eno' for him'. A good name, he suggested, was a man's best title. The popular press, which championed Thomas Ismay, celebrated his decision with the following ditty:

> He would not be a baronet,
> Not so his wishes ran,
> His mind on other things was set,
> He chose to be a man.[5]

The reason Thomas Ismay turned down the baronetcy was not, however, because he chose to be a man but because he hoped to be a lord. He felt insulted by the offer. A baronet was only a 'Sir' and did not count as a member of the peerage: he would effectively have remained a commoner. Thomas had previously put a good deal of effort into reviving the Ismay coat of arms, bestowed on the family in the reign of Edward I, and in 1891 Burke's Peerage informed him that the Ismay arms would be included in their next edition, along with the family motto: Be Mindful.[6] His family and friends urged him to swallow his pride, arguing that the acceptance of this honour 'would be no bar to further promotion (and that other shipowners were receiving recognition of their work etc.)',[7] but he was steadfast in his refusal. His son, James, suggested that he might have been too hasty: 'if the government had seen fit to offer you a suitable recognition of your great work, I should have been sorry if you had declined; for the bestowal of the honour I expected would have been thoroughly appreciated by the entire shipping community, and I am certain that everyone will feel that this can only be the first step towards the fitting elevation. Thinking over the other comparative cases, there does not seem to me any instance where a commoner, unless for special political reasons, has been given the higher rank without the intermediate step.' Bruce was more cagey on the matter: 'As you say,' he wrote to his father, 'whichever road one travels, one probably thinks it would have been better to have gone the other.'

His marriage was another of Thomas Ismay's success stories. Loyal, devoted, and equipped with a good business mind, Margaret Bruce made the perfect wife and their union offered her the chance to involve herself in his professional world. She kept a diary for fifty years and her daily entries, which document the progress of her husband's empire, reveal her worship of Thomas. Ismay Junior was raised in his father's light: he was the son of Midas.

Joseph Bruce Ismay was born on 12 December 1862. He had a sister, Mary, two years older, who died of scarlet fever aged eleven during a visit to her maternal grandfather. His younger brother, Henry, died aged two in 1866, when Bruce was four. Next in line was James, born when Bruce was five, followed by Ethel, who was three years younger than James. Margaret then gave birth in quick succession to two sets of twins: Ada and Dora in 1872, Bower and Charlotte in 1874. Thomas greeted his heirs in pairs, but apart from his second son, James, he did not much like his children and particularly disliked his eldest surviving child, who he saw as a mother's boy. Orphaned by his father's coldness and the mutual devotion of his parents, Bruce found himself sandwiched between two dead and much mourned siblings and succeeded by six others who fell into natural partnerships. He was the odd one out in a family of doubles.

Ismay's birthplace was Europe's western gateway to the New World. 'Busy, noisy, smoky, money-getting Liverpool,' as James Currie called it in 1804. 'This large, irregular, busy, opulent, corrupted town,' another visitor wrote ten years earlier. Known as the Marseilles of England, the great port was dominated by the traffic of people and goods: in 1840 it had sixteen docks; by 1900 there were forty. The city, whose skyline was a web of funnels and masts, was shaped by migration: some 9 million emigrants sailed from her harbours in the nineteenth century to start new lives in the United States, Canada and Australia, while immigrants from Ireland, Russia, Poland and Northern Europe poured into the city from the ships and trains. Liverpool's inhabitants knew the names of all the shipping lines and all the vessels; schoolboys could identify each company by its colours, they knew the tonnage, the length, the displacement figures of every new ship. Thousands of sailors and stevedores filled the quays, over-hauling the running gear, unloading the cargo, painting the boats. Talk in the Ismay household was dominated by steam and sail, speed and shipping routes, graving yards and gantries; Thomas Ismay's work seeped into every crevice of family life. Land, for young Ismay, was simply a place where ships came to fold their great white wings. The bay windows of the Ismays' three-storey house, Beech Lawn in the suburb of Waterloo, looked across Crosby beach onto the grime of the Mersey where the White Star ships would blow their sirens in tribute to their owner as they passed his residence. Ismay loved the mournful beauty of ships and his future, he knew, lay in crossing over to the other side. He grew up tethered to New York.

The first of his family to receive the education of a gentleman, aged eleven Ismay left his local school in New Brighton and went south to a fashionable preparatory school in Elstree, a pretty village on the outskirts of London. His years there, from 1874 to 1876, coincided with the brilliant headmastership of the lean, handsome Reverend Lancelot Sanderson, known to his pupils as 'the Guv'. Both the Reverend and his Irish wife, Katherine, were devoted to the welfare of their boys, each of whom was given a goodnight kiss by the ebullient 'Mrs Kitty', as Katherine Sanderson was affectionately called.

As well as a school of boys to look after, the Sandersons had thir-teen robust children of their own, of whom most were wayward and intelligent girls. The Sanderson tribe ran wild down the school

corridors in their homespun clothes, ragging and teasing one other
and anyone else they came across; they played hockey in the school
dining room with canes and a tennis ball, they staged elaborate theat-
ricals in the school hall, they pushed one another into the swimming
pool in the summer months and belted each other with knotted
towels. Monica, the eldest daughter, was remembered many years
after her death as having been a girl who 'sang in her cold bath'.[8] The
Reverend, whose health was frail, kept as far away from his unruly
spawn as he reasonably could, but the invigorating, unpretentious
atmosphere of his family suffused the school. In an age of decorum,
the intellectual and physical energy of the Sanderson way of life – as
well as the beauty of the daughters – made them hugely attractive
and Mrs Kitty's already vast dinner table groaned beneath the weight
of a constant trail of enthusiastic house guests who would prolong
meal times with philosophical and religious debate.

Visitors to Elstree tended to be men, like John Galsworthy and
Joseph Conrad, who had been picked up by one or other of the children on their various travels. Galsworthy and Conrad were initially
friends with Ted, the eldest of the Sanderson offspring, before being
embraced by the family as a whole. The unusual freedom of life at
Elstree was made apparent to Agnes, one of Ted's sisters, when she
visited the Galsworthy family home with its 'acres of red pile carpet'
and 'thousand gold chair legs'. The 'atmosphere of Victorian propri-
ety', Agnes remembered, 'the dumb immaculate servants, the low
flanneletty Galsworthian voices, killed all life in us'.[9] But the
Galsworthy home was Bohemian in comparison with the Ismay
household.

Elstree was a family school in every sense. Ted Sanderson would
succeed his father as headmaster in 1910, and he in turn would be
succeeded by his own son. For Ted, 'there was no other school in the
land to compare with Elstree. Where else was there such a fine view,
such a beautiful chapel, such a Classical tradition, such good
manners, such a manly, Christian tone?'[10] Most people agreed. In
1887, it was voted by the *Pall Mall Gazette* the country's best
prep school for boys. In *Sporting Pie*, a memoir of his schooldays,

F. B. Wilson writes that 'in the year of 1890 there can have been no private school in the world that was quite on a par with Elstree, if only by reason of the wonderful selection of masters whom the Rev Lancelot Sanderson had got together'.

It was a school in which a homesick boy might be happy, but Ismay was miserable at Elstree. His reserved nature made him particularly unsuited to the community spirit of boarding and to the continual joshing and mocking and testing of a boy's limits which went on from dawn to dusk. Even as a child, Ismay did not like the company of children; he was unpopular and lonely and separated from his adored mother, who had just given birth to her second set of twins. Raised in an authoritarian household, Ismay was uncomfortable around the domestic unruliness of the Sandersons. He liked his world ordered and controlled; as an adult, his visitors learned to arrive ten minutes earlier than their agreed time and wait until the clock struck the appointed hour before knocking on the door.

Ismay was a northerner in the south of England, a sea-gazer in a landlocked village. The only time he could return to the world of ships was in his school books, where he read about the HMS *Birkenhead*, the most famous shipwreck of the day, which had sunk off the Cape when his father was a teenager, carrying only enough lifeboats for the women and children. It was on board the *Birkenhead* that the 'age-old' law of women and children, known as the 'Birkenhead drill', was first used. Ismay learned how the band was reported to have played as the ship went down, how 'the roll of the drum called the soldiers to arms on the upper deck', how 'they stood, as if on parade, no man showing restlessness or fear, though the ship was every moment going down, down'.[11] He and schoolboys all over the country would recite Kipling's lines, 'stand and be still to the Birken'ead drill'.

Thomas Ismay sent Bruce to Elstree not because he valued its atmosphere of freedom, which he doubtless considered southern and soft, but because the school prepared its pupils for Harrow. Ismay, whose clear thinking was often remarked upon, passed the entrance exams in the autumn of 1876 and became a Harrovian in

January 1877 when he had just turned fifteen. The rigour of life at Harrow might have suited a boy of Ismay's temperament more than had the jumble of Elstree, but Ismay's misery seems to have increased. Again, he was under the tutelage of a legendary headmaster who served for other boys as an inspiration. Dr Montagu Butler raised his pupils in the Arthurian tradition of gallantry, honour and courtliness. Harrovians were marinated in the language of chivalry; this was a time in which the Victorian public schools promoted 'the knightly life once more', as J. H. Skrine, Warden of Glenalmond College, scathingly put it, with the attendant 'narrowness . . . pride of caste . . . soldier scorn of books and industry which is not of the open air, as war, the chase, the game'. English public schools were breeding a species of male so pointlessly romantic and impossibly arcane that they seemed, in the words of Second World War poet Keith Douglas, to be 'unicorns, almost'.

Ismay was in Bushell's House (as was Ted Sanderson, who went up to Harrow four years later) along with forty other boarders. He shared a room consisting of a table, two chairs, a washstand and two beds which folded into the wall during the day; he was expected to provide his own rugs, cups and cushions, either bringing them from home or buying them from other boys, and he soon learned that displaying pictures of your family was considered in bad taste. John Galsworthy, five years younger than Ismay, recalled his own time at Harrow – which had been a great success – as dominated by 'all sorts of unwritten rules of suppression. You must turn up your trousers; must not go out with your umbrella rolled. Your hat must be worn tilted forward; you must not walk more than two abreast till you reached a certain form.' You must 'not be enthusiastic about anything, except such a supreme matter as a drive over the pavilion at cricket, or run the whole length of the ground at football. You must not talk about yourself or your home people; and for any punishment you must assume complete indifference.'[12]

In his first term Ismay fagged for an older boy, which involved filling his bath, lighting his fires, running his messages, carrying his footballs, bringing him gravy cutlets and jugged hare from the village, and rushing to his side whenever he heard the call 'boy-oy-oy'. Other Harrow traditions included the folding of a bed to the wall with a boy still tucked up inside it and the 'House chor', at which the 'fezzes', or the football eleven, sat in state at a table with two candlesticks, a toasting fork and a racquet. Standing on the table, a candlestick in each hand, every boy in the house would take turns to sing, any sign of hesitation eliciting a prod from the toasting fork and a slap from the racquet.

That Ismay failed to distinguish himself academically at Harrow was inconsequential. He was sent there to become a gentleman and not a scholar, and in order to be a gentleman he needed to mix with the sons of other gentlemen. As Squire Brown in Thomas Hughes's *Tom Brown's Schooldays* (1857) puts it, on sending his own son to Rugby: 'I don't care a straw for Greek particles, or the digamma . . . If he'll only turn out a brave, helpful, truth-telling Englishman, and a gentleman, and a Christian, that's all I want.' Dr Thomas Arnold, Rugby's famous headmaster, claimed that what his school looked for in a boy was, 'first, religious and moral principle; secondly, gentlemanly conduct; third, intellectual ability', and the same was true for Harrow. Any learning that took place beyond an introduction to the ethics of leadership, fair play and self-control was incidental.

The public schools were in the business of developing 'character' rather than intellect, and character meant conformity, selflessness, and patriotism.[13] In his novel *The Harrovians* (1913), Arnold Lunn – a sportsman and adventurer whose father founded the Lunn Poly travel agency – praised the public schools for aiming 'at something higher than culture. They build up character and turn out manly, clean-living men that are the rock of empire. They teach boys something which is more important than the classics. They teach them to play the game. It does not matter what a man *knows*. It's precisely what he *is* that signifies.' There was no better way for a boy to learn how 'to play the game' than through sport. Sport formed 'character' and Ismay was at Harrow during the years that games, particularly football and cricket, became a cult. 'A truly chivalrous football player', the *Marlburian* declared in 1876, 'was never yet guilty of

lying, or deceit, or meanness, whether of word or action.' A school
athlete was a hero, an honour to his house, a proven leader and a
moral beacon. With his strapping build and impressive height, Ismay
excelled at all forms of sport, particularly tennis and shooting, and
yet his school records show that he joined no school team. Nor was
he elected a member of the Philathletic club, founded by the boys
themselves in 1853 and composed chiefly, though not entirely, of
prominent athletes. Members were elected, and the fact that a fellow
was a 'good sort' weighed when voting.

Being nouveau riche, Ismay could never be a 'good sort' any more
than he could have 'character'. Despite the growing respect shown by
'old money' towards trade, coupled with the appreciation that many
public school boys would go into business themselves, it was understood that earning a place in the world of commerce was not the
same thing as inheriting a place in society. The middle classes could
be educated alongside the upper classes, could imbibe their values
and even, given the right moral fibre, become gentlemen themselves;
but the stigma of trade nonetheless remained. Behaviour was held to
be an effect of breeding and every new boy at Harrow was welcomed
with the same question: 'Who's your father?' Those with 'heroic'
fathers, brothers or uncles, men who had excelled in sports or died
for their country, were accorded heroic status themselves, while those
like Ismay, who came from nowhere, were treated as nobodies.

Starting Harrow in the same term as Ismay was Alexander Arthur,
whose father was also a Liverpool shipowner. Perhaps the two new
boys were friends – doubtless their fathers were acquainted – but it
seems more likely that they sniffed one another out immediately and
kept a good distance. Ismay was in an impossible position: the father
he feared and respected at home was the object of derision at school.
Thomas Ismay was relatively uneducated and spoke with a Cumbrian
accent; his particular genius was of no value outside the dockyards.
While his parents boasted that Bruce was at a top public school, he
was being taught to feel ashamed of his home and of the family business he was expected to inherit. His father, meanwhile, had
ambivalent feelings about the expensive education he was providing

for his son. He was not putting Bruce through Harrow in order for the boy to gain notions of superiority or have ideas of his own; he was being primed to fill his father's shoes.

Were it not for his background, Ismay would have blended in with the others; he could never be accused of being either a swot or a wimp. In his history of the family, *The Ismay Line*, Wilton Oldham suggests that Bruce's unhappiness at school was due to his having inherited his mother's 'shy and sensitive' nature. Bruce, Oldham writes, learned to put on a 'façade of brusqueness to avoid appearing over-sensitive' which 'often caused people to dislike him, until they got to know him well'.[14] But few people, and certainly no one at Harrow, got to know him well; what friendships Ismay formed came later in life. He would have been a target for bullies but it is more likely that, as with many unhappy and conflicted children, Ismay himself was the bully. He had learned from his father the art of intimidation, which he now used in order to build a wall around himself. He would not have been a glamorous bully like Flashman, who terrorised the Rugby of *Tom Brown's Schooldays*; he is more likely to have been a silently threatening loner, someone who might slap a bed against the wall with a small boy tucked inside. Pupils kept away from Ismay, and mocked his lack of pedigree only when his back was turned.

For Galsworthy, the public school creed was as follows: 'I believe in my father, and his father, and his father's father, the makers and keepers of my estate, and I believe in myself and my son and my son's son. And I believe that we have made the country and shall keep the country what it is. And I believe in the Public Schools, especially the Public School I was at.'[15] What is striking about Ismay is that he believed neither in his father, nor in his son (who would go to Eton), nor in the public school he was at. Ismay never identified himself as a Harrovian, never nurtured a nostalgia for his alma mater's rituals, private language or house songs; never saw himself as belonging to a exclusive club of fellows who had more in common with one another than they had with the rest of society. He felt no connection to the school whatsoever.

In Ismay's same year and house was Horace Vachell, who later wrote a novel about Harrow called *The Hill: A Romance of Friendship*. Published in 1905 but set in the 1890s, *The Hill* gives us a good idea of what Harrow was like for someone like Ismay, and it is possible that Vachell's main character, a bully called Scaife who is the unpopular son of a Liverpool shipping magnate, was a portrait of Ismay himself. Vachell chose as his novel's epigraph 'Fellowship is heaven and the lack of it is hell'. The lines, by the pre-Raphelite William Morris, which continue 'fellowship is life and lack of fellowship is death: and the deeds that ye do upon the earth it is for fellowship's sake that ye do them', capture the Edwardian cult of belonging, and the fourteen-year-old Scaife, who does not belong to the heaven of fellowship, is appropriately known as 'the Demon'. Scaife's father, in the manner of Tom Brown's father and of Ismay's father too, tells him that 'I'm sending you to Harrow to study, not books nor games, but boys, who will be men when you are a man. And above all, study their weaknesses. Look for the flaws. Teach yourself to recognise at a glance the liar, the humbug, the fool, the egoist, and the mule. Make friends with as many as are likely to help you in after life, and don't forget that one enemy may inflict a greater injury than twenty friends can repair. Spend money freely; dress well, swim with the tide, not against it.'

Making friends, however, presented difficulties for Scaife because his father pronounced 'inestimable' 'inesteemable' and 'connoisseur' 'connysure'. 'Ah, the Scaifes!' the father of one boy says; 'A man I know dined with them last week. He reported everything *overdone*, except the food.' Socially suspect, Scaife is morally queasy: a boy from his background might have all the credentials for popularity, but he can never be *one of us*. Scaife was 'keen at games, popular in his house, clever at work – clever, indeed! Inasmuch as he never achieved more or less than was necessary – generous with his money, handsome and well-mannered, blessed, in fine, with so many gifts of the Gods, yet [he] lacked a soul.'

'One is reminded sometimes,' says the languid son of an old family, 'that the poor Demon is the son of a Liverpool merchant, bred in or about the Docks . . . One knows that family is not everything, but, other things being equal, it means refinement. The first of the Howards was a swineherd, I dare say, but generations of education, of association with the best, have turned them from swineherds into gentlemen, and it takes generations to do it.'[16]

On board the *Titanic* was another Harrovian, Tyrell William Cavendish, who was in the same year as Ismay's youngest brother, Bower. Tyrell Cavendish, who went down with the ship, was described in a letter to the *Uttoxeter Advertiser*, his local paper, as 'one of the heroes' who 'died as an English noble gentleman, unselfish and heroic to the last'. The paper did not report Ismay's survival, which was nothing if not a proclamation of selfhood as something distinct from fellowship.

Ismay left Harrow after eighteen months. It is unlikely that Thomas pandered to his son's unhappiness by removing him from the school; Bruce probably left because Harrow's work was complete. He could now enter the world a Harrovian and that is what counted. He was sent to complete his education in a tutoring establishment in Dinard, a fashionable French resort frequented by wealthy English and American tourists. Here he learned to play excellent tennis and his tutor, an English clergyman called the Reverend Edwards, predicted that Bruce 'will . . . be one of the leading men in the country; he has such a wonderful brain'. His abilities were best suited to the world of business, and when Ismay returned to England he went not to Oxford, as his brother James would do, but back to Liverpool where he started as an apprentice at Ismay, Imrie & Company.

In *The Ismay Line,* Wilton Oldham recounts how on Bruce's first day at work he left his hat and coat on his father's stand, as he had done throughout his childhood. 'Please inform the new office boy', Ismay Senior asked one of his clerks, 'that he is not to leave his hat and coat lying about in my office.' By means such as this, Oldham suggests, Thomas 'implanted' in Bruce 'the sense of inferiority' he then carried all his life. But what is striking about the tale – apart from the fact that Ismay Junior could not distinguish between the etiquette of work and that of home – is that he was so wounded by his father's insistence that he use the same cloakroom as his new colleagues that, in the words of Oldham, 'he rarely wore a coat again'.

At around the same time, he returned home from work early one evening and, without asking permission, took his father's favourite horse for a gallop along Crosby Sands. When the animal broke his leg and had to be shot, Thomas's rage was such that Bruce 'never rode again'. This pattern was to be repeated throughout his life: when his own eldest son died as a baby, Ismay found contact with his next three children difficult; when his second son, Tom, caught polio, he further distanced himself – 'Bruce possessed that curious trait which some people have,' wrote Oldham, 'in that he shrank from anyone who was not physically perfect and after this his attitude to Tom was tinged with this involuntary repugnance.' When the *Titanic* was sinking, Ismay could not watch, and he then never went to New York again.

Oldham, who is an apologist for Bruce Ismay – 'I have always felt that he was the most misunderstood and misjudged character of the early part of the century' – edges around the tricky relationship between father and son.[17] Bruce, he writes, 'was devoted to both his parents and his mother loved him too; but his father found him difficult'. If Bruce was 'brusque and arrogant', filled with 'destructive criticism' and a 'biting sarcasm', it was because he had been broken by his father's 'constant humiliations'. There was, Oldham says, 'a feeling of constraint between them owing to Thomas Ismay's unconscious jealousy of [Bruce]'. The 'friction' between them was such that they could not occupy the same house or be in the office at the same time as one another. Oldham's comments are the result of conversations with Ismay's widow, Florence, then in her nineties; the specific reasons why Thomas might have found Bruce difficult are not discussed, and nor does Oldham offer reasons for the mutual 'friction' or the 'unconscious jealousy'. What is clear is that Thomas Ismay, who generally found people easy to deal with because they did what he told them to do, was irritated by Bruce, who did not. Yet Bruce was the only one of the three sons, including Thomas Ismay's favourite, James, to share his father's love of ships and shipping, and the only one prepared to devote himself to the White Star Line. Ismay Senior was a tycoon who dreamed of heading a dynasty, and Bruce was the means by which he could achieve his ambition. James, who had no interest in ships, proved himself an excellent landlord and farmer, while Bower – Ismay's favourite brother – grew into an Edwardian dandy who squandered his father's money on racehorses. The only explanation Oldham gives of the loathing between Thomas and Bruce is that Bruce 'was quick to learn and when asked his opinion would state it clearly and forthrightly; Thomas Ismay liked to consider a problem from all angles before reaching a decision, and so resented his quick-thinking son'. Bruce was, according to those who worked for him, dogmatic and dictatorial; he would brook no argument and his insistence on punctuality verged on the fanatical. The problem for Thomas was that Bruce was the same as him and different, and he feared both aspects of his character. He wanted his eldest son to be his mirror image but not to occupy the same space. Thomas provided him with palatial homes, fleets of servants and an upper-class education, but then resented him for having it easy. Bruce grew up knowing that he was not himself but a failed version of someone else; he was never to forget that he was an inferior model, an imperfect copy.[18]

In 1877, when Bruce was away at school, Thomas Ismay bought a house surrounded by 390 acres of melancholy, dank land in Thurstaston on the western Wirral, overlooking the sandbanks of the River Dee and twelve miles by ferry from Liverpool. The Ismays' former home of Beech Lawn was in a suburb inhabited by sea captains and ship's officers. Captain E. J. Smith had at one time lived around the corner and Joseph Bell, the *Titanic*'s Chief Engineer, lived down the road, as would, at various points, Captain Rostron of the *Carpathia* and Second Officer Charles Lightoller. Now that Thomas Ismay had become rich, he needed to set himself apart from his employees. He needed his own Xanadu.

Throughout the second half of the nineteenth century, the
Liverpool suburbs were filling up with merchant palaces. 'Crowds of
comfortable and luxurious villas', wrote a journalist in 1873, 'besprin-
kle the country for miles round Liverpool, inhabited by shipowners,
ship-insurers, corn merchants, cotton brokers, emigrant agents,
&tc, &tc, men with "one foot on sea, and one on shore"'. One such
villa was Broughton Hall, the home of Gustavus Schwabe. A Gothic
Revival mansion (today a convent), Broughton was built for
Schwabe in 1859 and it was here in 1869 that the agreement had been
made between Thomas Ismay and William Imrie to resurrect the
White Star Line. Thomas Ismay's purchase of the land at Thurstaston
was a demonstration of how successful the previous seven years had
been. Imrie, meanwhile, had moved into a grand pile called
Holmstead on the North Mossley Hill Road. A patron of the arts,
Imrie filled his house (now also a convent) with Pre-Raphaelite
paintings, including his collection of works by Evelyn De Morgan.
Holmstead was the epitome of modernity and discernment:
Rossetti's *Dante's Dream* could be found above the fire in the library
next to a stained-glass window by Morris and Co., and Edward
Burne-Jones's *The Tree of Forgiveness* was displayed upon William
Morris wallpaper in the music room where Imrie and his wife would
host singing evenings. Frederick Leyland, another self-made
Liverpool shipowner who patronised contemporary artists, entertained Whistler in his own new manor house, Speke Hall, eight
miles out of Liverpool. Thomas Ismay's house would prove him likewise to be a man of taste: his home would be the perfect marriage of
money and art. He would appear to all the world the image of
fulfilment.

Ismay Senior liked making things – in the garden of Beech Lawn
he had built a grotto consisting of large rocks faced with mirrors
which surrounded a sunken well and iron spiral staircase – and he
now decided to demolish the existing house in Thurstaston, built
only twelve years earlier, and start again. He would ask the most
fashionable architect of the day to design him a mansion that would
overshadow all other mansions. Not one nail would go into its

construction: Thomas wanted the building held together by brass screws alone. No reason is recorded for this particular eccentricity but it is possibly because the house would then resemble a ship, where steel plates are bolted together by rivets. To gather ideas for his new project, he and Margaret began a tour of English manors and in January 1882 they visited Adcote in Shrewsbury, designed by Richard Norman Shaw for the Darby family, the industrialists who built Ironbridge. Shaw's style was a combination of 'merrie Englande' and Gothic Revival; an admirer of Pugin, Shaw designed vernacular buildings typified by severe outlooks and soaring, twisted chimneys; other signature touches were half-timbered bay windows, inglenooks, stained glass and high ceilings. Thomas Ismay liked his style and asked Shaw to give him something akin to Adcote, only bigger and better. He wanted a modern house which impersonated an ancient house, something which, as Charles Ryder would describe Brideshead, looked as though it had grown silently with the centuries, catching and keeping the best of each generation.

Shaw, who had a reputation for building stately homes for businessmen (his customers would include a number of shipowners), also had shipping connections: his brother was a partner in Shaw Savill, a company specialising in runs to New Zealand, and during the construction of the Ismay house, White Star made a deal to provide the ships and crew for Shaw Savill's New Zealand trade if they came up with the passengers.

Shaw's legacy can be seen in London's New Scotland Yard, the striped White Star offices in Liverpool, and the Tudorbethan houses of the stockbrocker belt. He built sturdy mansions for sturdy men and Dawpool, the mansion he built for Thomas Ismay, was, in the words of Shaw's biographer Andrew Saint, 'a monster perched on the headland'.[19] Again and again, Dawpool, whose front alone measured 250 feet, is described in terms of monstrosity. It was built to be magnificent but, unlike the White Star liners, Dawpool's treatment of its guests was indiscriminate. No one was comfortable; Dawpool was an exercise in strength and severity rather than in home-making.

Florid at sea, Ismay Senior was austere on land. Dawpool's immense mass, Saint writes, 'palliated only by dark ivy patches clasping at the ruddy sandstone walling, started out from the barren heath, amidst outcrop, gravel, furze, heather and bracken'. The mood was 'alternately fearsome and ghostly, lowering in shadow and rainfall, bleaching balefully in sunlight, in accord with the rich but chilly red of the Wirral sandstone from which it was fashioned'.[20] Thomas had no interest in cultivating land, and the consequent lack of trees, gardens (he disliked flowers) and parkland increased the general aspect of brutalism.

Dawpool's interior was equally sobering. The dining room, says Saint, was 'reminiscent of those pictures by Orchardson in which the grandee and his bored wife dine in desperate solitude'. Shaw had 'over-relied' throughout on set-piece effects such as his trademark part-Tudor and part-Classic inglenooks.[21] An enormous, palm-filled and panelled picture room, whose domed, glazed ceiling ('an eyesore', according to Margaret Ismay) appeared to be supported by a colossal chimneypiece constructed from two enormous church organs, served as a gallery for Thomas Ismay's art collection, which included Rossetti's *The Loving Cup*. Keen to compete with collectors of contemporary art, Ismay Senior also had his portrait painted by John Everett Millais.

Costing £50,000 (£3.5 million in today's money), the house took four years to build and was completed in 1886. It was not meant to be a working building in the heart of a great estate, but nor was it meant to be a family home. Dawpool was a show house, a stage set. The family, when they moved in with a newly acquired retinue of twenty-two indoor and ten outdoor servants, loathed it. Because there was no central heating and Thomas didn't allow fires in the bedrooms until teatime, the house remained freezing for most of the year. When they were at last lit, the fires smoked so badly that, regardless of the weather, the windows had to be opened in order to release the fug. In these chilly rooms the unmarried Ismay daughters, forbidden by their father either to speak at table or to read the newspapers, sat sewing with their mother.

So pleased was he with Dawpool that in 1897 Thomas Ismay asked Shaw to design the interior of the White Star's new flagship, named *Oceanic* after their first liner. This second *Oceanic* was to be the largest and most luxurious ship in the world and the first to exceed the tonnage of Brunel's *Great Eastern* of 1860. Money was no object, and so Shaw duly equipped her with gold-plated light fittings, marble lavatories, Queen Anne mirrors, Adam fireplaces, Ionic pillars, a domed ceiling decorated with allegorical figures representing Great Britain, the United States, Liverpool and New York, an oak-panelled dining room awash with gold, further panels carved with fruit and flowers in the style of Grinling Gibbons and a bacchanalian procession. There was a magnificent 'Turkey red' central staircase, and a smoking room lined in highly embossed leather gilt, with a carved mahogany frieze depicting sea nymphs. The sea nymph theme continued in stained-glass sliding shutters, while the ceiling formed a double dome decorated with scenes from the life of Columbus, and marble Italian figures posed in the corner niches. The *Oceanic*'s passengers, one commentator said, might be in Haddon Hall, the famous Elizabethan manor in Derbyshire; this was the grand country-house style which would increasingly characterise the interiors of all the White Star liners. Only in third-class, with its narrow bunks and undisguised pipes and girders, did the ship still resemble a ship.

The problem for Thomas Ismay was that while he did not want to share his son's company, he did want him involved in the family firm. To achieve both ends, the year after moving to Dawpool he sent Bruce, now aged twenty-one, to New Zealand aboard the newest White Star liner, the *Doric*. This was a luxury cruise rather than an adventure of the sort Thomas had enjoyed on the *Charles Jackson*. There were to be no drenchings or drinking songs, there was no joshing with the crew; the point of the journey was not to test Ismay's mettle as a seaman but for him to see how the new

service was working, and he was treated throughout with the
wariness and respect due to the son of the owner. On his return
to England nine months later, the position of manager of the
White Star Line agency in New York became vacant and to keep
him further out of his way, Thomas arranged that Bruce be given
the job.

Three weeks after Ismay took up his new post he was confronted
by his first maritime disaster. Two White Star ships, the *Celtic* and
the *Britannic*, one on her way to Liverpool and the other on her way
back, collided in poor weather and were badly damaged. The passengers panicked and the Captain was forced to use his gun to maintain
order. The wounded liners limped back to New York in tandem,
flashing lights and firing guns at one-minute intervals. An inquiry
was held and the Captains of both ships were severely reprimanded
for travelling at excessive speeds. Ismay, the representative of the
company, was expected to issue the US press with information. It
was a job to which he was ill-suited: Ismay loathed publicity and
had never mastered the art of public speaking or self-presentation.
He had a brusque, abrupt and imperious manner, and journalists –
to whom he was always rude – saw him as arrogant, charmless and
disdainful. But most striking was his inability to give a straight reply
to a plain question. Ismay became famous amongst reporters on
both sides of the pond for what the *Financier* called his 'enigmatic'
non-answers. In its 'Shipping Notes', the paper described Ismay's
responses in an interview as 'the luminous fog' which 'issued from
his mouth'. But no one disliked Bruce Ismay as much as the press
baron and man of the people, William Randolph Hearst (portrayed
as Citizen Kane in the film by Orson Welles). The combination of
Hearst's hatred and Ismay's luminous fog would result in his
undoing.

Aside from unpopularity with the press, all we know of Ismay's
activities during his two years in New York is that they caused his
parents concern. In Wilton Oldham's words, he 'painted the town
very red indeed', which probably means that he behaved much as
any rich, handsome, unattached twenty-two-year-old male would do

in a city 4,000 miles away from his oppressive father. The evidence suggests that Ismay was happy; he socialised, he made friends; he was away from English snobbery in a city where every man was a potential tycoon. He went to dances and concerts, and he indulged his love of expensive shoes and suits – Ismay, whose 'clothes', in the words of an employee, 'were always perfect and his shoes a dream', was described by the *Boston Globe* as 'somewhat fond of club life and one of the best-dressed men in England'.[22] When Thomas offered him a partnership in Ismay, Imrie and Co., the acceptance of which would require him to return to Liverpool, Bruce – to his father's fury – turned it down.

It was in New York that Ismay met Harold Sanderson. No relation of the Sandersons of Elstree, Harold, who was born in 1859, came from old Yorkshire stock and was the eldest son of Richard Sanderson, who ran a shipping agency in Birkenhead for the Wilson Line of Hull. In the early 1880s, Richard Sanderson had left England to represent the Wilson Line in New York. Here, with Harold and his younger brothers, he founded his own shipping firm of Sanderson and Sons. Six foot two and as lean as a greyhound, Harold Sanderson was 'courteous and considerate to a degree hardly attainable by others', with a reputation for honesty in business.[23] United by their youth, nationality, professions and physiques, Sanderson and Ismay became friends.

Harold Sanderson had married his wife, Maud Blood, in New York in 1885 and it was his example that Ismay was following when he gave up his life of pleasure in 1887 and proposed to Florence Schieffelin, 'a charming girl with real brown hair, beautiful eyes and a singularly winsome manner'.[24] Maud Blood, to whom Sanderson was devoted, was from a cultured and well-travelled family and had been educated in Germany, Switzerland and Italy. Florence Schieffelin, the eighteen-year-old daughter of one of New York's oldest families, came from similar stock. The Schieffelins originated from Nordlington, near Nuremberg; a forebear had been a distinguished artist and pupil of Albrecht Dürer. Florence's father, George, was a graduate of Columbia and a prominent New York attorney;

her mother, Margaret, was the granddaughter of John Ferris Delaplaine, a wealthy New York shipping merchant. Within weeks of setting eyes on her, during a Thanksgiving Day party at the Tuxedo Club in New Jersey, Ismay asked Florence to marry him. Neither father approved of the match and George Schieffelin, who doted on his daughter, suggested that the couple wait a year. When he finally gave his consent, it was on the grounds that Ismay promised always to keep Florence in America. The arrangement suited Ismay perfectly. They were married in December 1888 in the Church of the Heavenly Rest on Fifth Avenue, before what the *New York Times* called 'a fashionable assemblage'. The bride, celebrated as the 'belle of the city', wore lace and diamonds and carried a bouquet of lilies-of-the-valley. The papers described Ismay as 'a fine specimen of young manhood, being about twenty-six years of age, tall and graceful. He has, at this early age, achieved what might be deemed an enviable position in the marine world where he is widely known, popular and a great favourite, as he is also in the social world of New York.' The wedding breakfast was held at the Schieffelin family home on East Forty-Ninth Street, after which the Ismays moved into a house at 444 Madison Avenue.

The groom's family did not attend the ceremony; Thomas Ismay was too busy with Lord Hartington's Commission on the administration of the Army and Navy. A proud man, Ismay Senior nurtured doubts about the clan Bruce was marrying into: the Schieffelins were not only foreigners but also upper class, with all the attendant notions of superiority. He was equally uncomfortable when, in 1892, James married Lady Margaret Seymour, the eldest daughter of the Marquis of Hertford. He did not want it thought that he tried to advance his family by engineering 'good' marriages; the sons of Thomas Ismay did not need to marry well in order to prove themselves gentlemen. He and Margaret did not meet Florence until after the wedding, when Bruce brought his bride to England. Their arrival at Dawpool was then celebrated as befits the son of the lord of the manor, with the ringing of bells, fireworks and a dinner for one thousand of the residents of Thurstaston.

A daughter, Margaret, was born to the newlyweds in late 1889 after which, in September 1890, Thomas, who wanted to retire from the company but remain involved as its chairman, again suggested that Bruce become a partner, along with James who was just down from Oxford. The offer was once more contingent on Ismay's return to England. Should he turn it down, the position would go to James. Ismay, who resented the idea of being overtaken by a preferred brother with no interest in ships and none of his own experience – Bruce had now worked for the White Star Line for ten years – had little choice. He was caught between the demands of his father and the promise he had made to his father-in-law to keep Florence in America, and Thomas doubtless enjoyed playing the trump card in the power struggle between the two patriarchs. George Schieffelin reluctantly released his son-in-law from his bond, and in January 1891 he saw his daughter, his granddaughter, and his six-month-old grandson, Henry, off to Liverpool on board the White Star liner, the *Teutonic*.

The crossing was a disaster. Florence was seasick, baby Henry became ill; Florence wanted to nurse him but Ismay insisted on using an inexperienced maid. The baby's illness worsened, and soon after they landed, Henry died at Dawpool. The event brought to an end what marital happiness the Ismays had briefly known. Bruce, who would never recover from his young son's death, now fixed his affections on his daughter. Florence had sacrificed her child and her country to start a new life in a drab Liverpool suburb with a man she no longer recognised.

Margaret Ismay suggested that the young family stay at Dawpool until they found a suitable home of their own, but it was an idea Thomas could not countenance. His life was ordered with military precision; if he saw a fallen leaf on the drive when he left for work in the morning he would put a stone on it; should it still be there when he returned in the evening he demanded of the gardeners what they had been doing all day. The thought of having at the heart of his empire a taciturn son, a tearful American and an unpredictable two-year-old granddaughter was intolerable, so he rented for them what

Florence called a 'horrid house' several miles away, and threw in the use of a servant to help with the cooking.

Bruce immersed himself in work, determined that his father would have no grounds for criticising him. There were no more weekend jaunts for the couple, no more dances. Pleasures that he and Florence had once enjoyed together, such as the theatre, he now preferred to do alone, slipping out to a matinee during a quiet afternoon. Unlike his father, who had shared all his business concerns with his wife, Ismay shared nothing with Florence. He would think his problems through in solitude during long walks, or on the top deck of the trams in which he would travel around and around the city. He stuck to a clockwork routine, leaving the house at the same time every morning and returning at the same time every night, bringing with him what Florence referred to as his 'dreaded dispatch box'. The only outings with his wife were now dinners at Dawpool, which Florence also thought horrid. Her presence there, however, added a lightness of touch and Ismay's four sisters adored her. Regardless of Thomas's expectation that women remain mute at the table and disengage themselves from the wider world, Florence, swallowing her homesickness and dressed in the latest fashions, chattered away in her American accent about her family and New York. Ismay, meanwhile, turned inwards. He grew to hate noise and particularly parties, but there was nothing he now disliked so much as a wedding.

It was Ismay's suggestion that Harold Sanderson leave Sanderson and Sons in the hands of his younger brothers and take on the job of general manager of the White Star Line in Liverpool. With Sanderson on board, Ismay would have a colleague to whom he could relate while Florence would have, in Maud, a compatriot as a neighbour. The job would be a step up the ladder for Sanderson, and Thomas Ismay, impressed that his son was making sensible decisions, agreed to take him on, making Harold Sanderson the White Star Line's first major outside appointment. So in 1894 the Sanderson family, along with their six-foot-tall black nanny, arrived in Liverpool and moved into a sandstone house called Holmfield, which was next door to Sandheys, the late Georgian house in which the Ismays now lived in the suburb of Mossley Hill. Holmfield was soon filled with Maud's childhood friends from the continent, and the Sanderson children grew up in a 'babel of languages'.[25] To escape from his own silent household, Ismay took Sanderson for epic cycling trips, the two men sometimes covering up to seventy-five miles a day.

The culmination of Florence Ismay's unhappiness came in the summer of 1900 when, heavily pregnant, she went into labour with her fifth child (she and Bruce now had Margaret, Thomas and Evelyn) during a stay with the recently widowed Margaret at Dawpool. Not wanting the bother of a birth in her house, Margaret sent Florence back to Sandheys in a horse and carriage. During the journey back, a baby girl was stillborn and Florence never forgave her mother-in-law; nor did she forgive Bruce for allowing such a thing to happen.

Florence's wretchedness is vividly described by her granddaughter, Pauline Matarasso, in a memoir of her childhood, *A Voyage Closed and Done*:

> The Mossley Hill years were not happy ones for my grandmother: conventional, unimaginative but fun-loving, she was made wretched by what passed for intimacy with her husband and by the tedium of provincial life . . . Florence retained a lasting nostalgia for her childhood home, which translated itself into a refusal to eat anything she hadn't eaten in her father's house: fifty years later this still ruled out a number of staples, notably sausages. Brought up to compliance and respectful of all conventions, she released the bully in her husband who took pleasure in snubbing her at dinner parties, leaving her floundering and the guests embarrassed.[26]

The bullied son had become the bullying husband and father, and Florence 'slowly created a life for herself in the large spaces which her husband left her'. Ismay gave her an allowance with which she ran the home and garden and looked after the needs of the children and servants without having to bother him. She arranged the household, Florence later told Wilton Oldham, 'to revolve around Bruce; everything was done for his comfort and convenience, his tastes and preferences were studied meticulously and put before all others'. Because Ismay liked cold turkey, cold turkey was served every night; because Ismay did not like the noise of children, except for that of Margaret, they were housed in a specially built wing, separated from their parents by two green baize doors. In an unsigned document written in 1936, a family friend made an attempt to describe and explain Ismay. 'Treated with undue severity in his youth, he makes no allowance for its follies and ignorances. Nor has he learned that repression is not the way to encourage development in the young.' His 'otherwise fine character is spoiled by bursts of irritability and unreasonableness' which he directs against 'those of whom he is fondest'. Had Ismay's behaviour, the writer concludes, 'been commented on by the sweetest and best of women in their early married life, a good deal of suffering might have been avoided'.

Florence collected antique furniture and nurtured her brood, especially Tom who, after contracting polio as a baby, had become 'the butt of his father's sarcasm'.[27] In 1905 she purchased her first motor car, and owing to Ismay's dislike of driving, she began to enjoy motoring holidays alone. To provide some companionship for his daughter, George Schieffelin had sent over her rebellious, voluble, horse-loving sister, Constance. 'Con' was liked by everyone but her charms were particularly admired by Ismay's youngest brother, Bower, the only male twin of the two sets in the Ismay family. Before either had time to discover they had nothing in common beyond being related already, Bower and Constance were married. It was 1900 and a family of doubles had doubled itself even further.

Following a series of heart attacks, Thomas Ismay had died aged sixty-two in November 1899, one year before the reign of Queen Victoria came to its end. He lived just long enough to see, in January, the launch of the *Oceanic*, whose magnificent interior had been designed by Shaw. The 'crack' liner that was to be a monument to himself and his firm became instead his memorial. Visitors from Britain, Germany and America came to watch the *Oceanic* launched, and a special train brought sightseers from all over Ireland to attend the occasion at Harland & Wolff. When she then steamed up the Mersey, Ismay Senior, now seriously ill, went out in the tender to meet her; she was the finest vessel he had ever seen. The *Oceanic* was also admired by the Americans, especially the Wall Street monarch, J. Pierpont Morgan.

Thomas Ismay had created one of the world's most profitable shipping lines; his fortune was equivalent to $40 million and his weight in the shipping world was such that his inconsolable widow received a message of condolence from the Kaiser. 'There is terrible, terrible blank in the house,' Margaret wrote in her diary following her husband's death. 'Shall we ever be able to live without him. All my life was centred in him, and as the time goes on it can only get worse for me.'[28] On the day of Thomas Ismay's funeral, the city of Liverpool went into mourning with flags flying at half-mast. His career was compared by one journalist to a 'staircase ascending upward, straight, regular, well ordered, firm in its setting, perfect in its surroundings',[29] and his gravestone in the churchyard at Thurstaston village was inscribed with the legend: 'Great thoughts, great feelings came to him like instincts unawares.' It is a curiously pertinent inscription, for the son as much as for the father.

The first decision that Bruce Ismay made as the new head of the White Star Line was to sell it. In 1901 he was approached by J. Pierpont Morgan who, having arranged the mergers that formed General Electric, Northern Securities and the Steel Corporation, was now looking to monopolise the North Atlantic trade by bringing the various American and European steamships under the ownership of one monster trust. The British shipping industry, which had thrived during the Boer War, was currently crippled by competition; should Morgan control the shipping lines, he could fix the fares at a handsome profit. He therefore offered the White Star shareholders ten times the value of the line's earnings for 1900, a particularly lucrative year. Ismay insisted on a further $7 million in cash and Morgan eventually bought White Star for $35 million. The ships would continue to sail under a British flag and to employ a British crew, but they would be owned by Americans. William Imrie, James Ismay and a third partner,

W. S. Graves, would retire and Ismay continue as managing director and chairman of the company, with Harold Sanderson at his side.

In December 1902, White Star joined the American Line (formally the British Inman Line), the Red Star Line, the Dominion, the Atlantic Transport and the Leyland Line in the International Mercantile Marine, known as the IMM, and Ismay, Imrie & Co. became, as Bruce Ismay put it, 'a dead letter'.[30] James Ismay, whose wife, Lady Margaret Seymour, had recently died in childbirth, was only too pleased to throw in the towel and he moved his young family down to Dorset where he became a successful and much-loved country squire. He replaced his tenants' run-down cottages with large black and white timbered houses in the style of Norman Shaw, and farmed pigs, whose lard and bacon were supplied to the White Star liners. Bower was interested only in horses and so long as his hobby was funded, he did not much care who owned the White Star Line. The female members of the Ismay family were less happy about the turn in events. 'This ends the White Star Line,' Margaret wrote in her diary, 'in which so much interest, thought, and care was bestowed and which was my dearest one's life's work.' Because a clause in Thomas Ismay's will instructed that none of his daughters should invest in any other shipping company, the sale of the White Star Line meant that Ethel, Ada, Dora and Charlotte were each forced to relinquish their inheritance.

Both taciturn sons of rich fathers, J. Bruce Ismay and J. Pierpont Morgan were cut from the same cloth. 'Money Talks but Morgan doesn't', wrote the *Tribune* after a dinner held in Morgan's honour; his study contained a plaque which read '*Pense moult, Parle peu, Ecrit rien*' (Think a lot, Say little, Write nothing). Bruce Ismay, wrote a reporter in the *Northern Whig* in February 1904, 'is one of the silent ones of the earth. Whether he be a genius or not, one may be sure that garrulity will never be his undoing.' The millionaires presented themselves as two halves in a marriage of equals, and while Ismay insisted that the name of the IMM be absent from bills of lading of the White Star Line and that the White Star offices occupy an independent building of their own, it was clear that Thomas Ismay's company had become one of Morgan's costly possessions, to be counted along with his fabulous yacht, the *Corsair*, his famous library, and his vast collection of art and gems.[31] Morgan, according to his British partner, Clinton Dawkins, was a 'physical and intellectual giant', a man with 'something Titanic about him'.[32] Compared by a Yale professor to Alexander the Great, and described by B. C. Forbes, founder of *Forbes* magazine, as 'the financial Moses of the New World', Morgan was an imperious, intolerant, irascible robber baron, the 'boss croupier' of Wall Street who subjected the country's entire economy to the 'psychopathology of his will'. He now controlled not only the American railroads, but also his own fleet of luxury liners to ferry him across the Atlantic. It was Morgan rather than Ismay who was henceforth treated on board White Star ships as the owner; before the *Titanic* was even built, Morgan selected from the plans which suite of rooms would be his.

Morgan's new combine initially received a positive reception in the States. The inflation of the US merchant marine would give the country 'a position of pre-eminence such as it has not enjoyed since the decadence of shipbuilding' after the civil war.[33] Morgan believed he was also doing the British 'a good turn' and expected to be greeted 'with open arms' when he visited London later that year; instead he found himself 'everywhere cold-shouldered, having been suspected of filching our mercantile ships'.[34] The British, who had been sanguine about Morgan putting their other shipping lines into his trouser pocket, were distressed by the loss of White Star. What, the public asked, would happen in the event of war if the nation's finest ships were no longer owned by the nation? Ismay's actions had 'virtually cede[d] to the United States the control of the North Atlantic shipping business'.[35] An article in the shipping journal, *Fairplay*, expressed the general concern:

> What many people do feel is the keenest regret that such a magnificent line as the White Star, not to mention the other great lines associated with it, aggregating nearly one million tons of our best shipping, should have passed from British to American ownership, and from British to American control. The Combine fleets are American to the backbone; Americans found the capital, and it is Americans who

appoint and pay the managers to this side. It is nothing but mere pretence to say that through the technical wording of the Company's Act they are in any sense British, through this technicality they are allowed to fly the British flag, a fact most people regard as a public scandal.[36]

Berlin's *National Zeitung* smugly reported that 'the blow to England is all the greater since the German companies have been able to keep out of the trust and maintain their independence'.[37] Ismay's understanding was that a partnership with the House of Morgan had been unavoidable. Had Morgan not bought up the British shipping companies, the Americans, as the British Ambassador in Washington put it, would have formed an 'avowedly hostile combine to run our ships off the Atlantic and squeeze them . . . out of US ports'. Before long, the Americans began to regard the IMM as a British firm, a giant White Star Line, while the British believed that the White Star Line had secretly become American.

To prevent Cunard from joining the International Mercantile Marine, the British government granted them a subsidy of £2 million to build the *Lusitania* and the *Mauretania*, both of which were to be used as naval ships should the circumstances arise. The terms of the subsidy were that 'under no circumstances shall the management of the company be in the hands of, or the shares of the company held by, other than British subjects'. The *Lusitania* and *Mauretania* were a huge success; at 31,000 tons and equipped with the innovative Parson's steam turbine, they were not only the largest but also the fastest ships afloat – the *Mauretania* would hold the Blue Riband for over a decade; the experience of travelling in her was, for Henry James, as if 'carried in a gigantic grandmother's bosom and the gentle giantess had made but one mighty stride of it from land to land'.

The IMM, which was increasingly accompanied by the term 'ill-fated', would prove Morgan's only disaster. He controlled 136 ships and 45 routes between Europe and North America, but was operating at a loss. Cunard's independence had effectively handicapped the combine, which was forced to spend its profits on the construction of competing ships. 'The ocean', concluded the *Wall Street Journal*, 'was too big for the old man.' 'What threatens to swamp us,' wrote Clinton Dawkins, one of Morgan's British partners, to Charles Steele, one of his American partners, 'is this monstrous indebtedness for shipbuilding, and I don't feel satisfied that we are not putting more big ships into the Atlantic than it can bear.'[38] When share prices plummeted in 1904, Morgan decided to replace the ailing Clement A. Griscom as the trust's president. His first choice had been the brilliant Albert Ballin, head of the Hamburg-America Line, but Ballin turned it down and Morgan turned to Ismay as second best. He 'did not mind losing money', he explained to the White Star chairman, 'but he did object to doing so owing to poor organisation'. Morgan's hope was that the image of a young English captain of industry at the helm would push up the value of the company at the same time as quelling any alarm felt by the British at the 'Morganisation' of their favourite shipping line. Character, Morgan believed, comes 'before anything else. Money cannot buy it,'[39] and Ismay, he explained to his partners, not only possessed character but was 'the only man who could straighten matters out'.[40] Seven months earlier, Morgan had thought otherwise, arguing that an English president of the IMM would be an impossible drawback. In an interview with the *New York Mail* on 28 January 1904, Clement Griscom expressed his complete faith in Ismay as his replacement. 'He has been trained in a splendid school – a school where one-man management has been taught and practiced . . . Mr Ismay would not accept the presidency unless he be given the one-man power which has made the White Star Line so successful.' Ismay would of course, the American papers agreed, be making his home in New York. It would be impossible – as well as insulting – to run a vast American company from 'the other side'.

As always when making a decision, rather than turn to his wife for advice Ismay approached his mother and Sanderson. 'The idea of spending so much of my time in America is not congenial to me,' he wrote to Sanderson from New York. 'But if I could arrange to limit it to say, three months, it would not be so bad . . . I think you know that if I had considered my own inclination and feelings absolutely I should, in all probability, have resigned ere this, but I am trying to look at the matter from a general point of view.' 'I think your inclination is to accept,' Margaret Ismay told her son, 'and I do not wonder at it. For it is well known what exceptional power you have . . . You have a great deal of your life before you, and I hope you may be spared to bring the great concern with which you are so interested to a successful issue.'[41]

Ismay's inclination, however, was not to accept: 'If I only considered *myself*, I would decline the responsibility.'[42] He knew the position was humiliating, added to which he had grown to dislike New York, no doubt due in part to animosities from his wife's family. The presidency would also require a good deal of entertaining and public posturing, neither of which activities he enjoyed, but there was one further factor in Ismay's reluctance to take up Morgan's offer. His selling of the White Star Line was the start of the process by which he hoped to divest himself of responsibility in the shipping world and prepare for what he called, in a letter to the Head of the British Committee of the IMM, 'a life of ease and enjoyment'.[43] Ismay was forty-two and had, as his mother said, a great deal of his life before him, but he was uncertain that he wanted to spend it fulfilling his father's ambitions. He had other interests; he liked hunting and fishing and had hoped to be selected as a Unionist Member of Parliament for the constituency of Ludlow in Shropshire, but failed to get the support of a 'certain section of the Conservative party'.[44] J. P. Morgan, now seventy-five, was also feeling his age and wanted to spend more time with his paintings. Morgan was handing the baton over to his son; Ismay had brought to an end his own father's dream of a dynasty. Nonetheless, 'it is a difficult proposition', he wrote to Sanderson, 'and I intend going slow, and giving the matter the most earnest and careful consideration. It is easy to jump in, but it would be difficult, if not impossible, to climb out.'[45]

A cartoon in the *New York Globe* pictured John Bull, his head emerging from the Atlantic, devouring the IMM in the form of a ship. 'He's swallowed it!' says America, but the truth is that when Ismay jumped, it was J. Pierpont Morgan who swallowed him whole.

"He's Swallowed It. Be Gosh!"

His appointment, noted the *Glasgow Evening Times*, 'was conso-
nant with the whole of the Ismay family history. An Ismay never
goes back.' And Ismay did not return to live in New York, a slight
that the city would find hard to forgive. Instead, as he patiently
explained to journalists wondering where he would now be residing,
the headquarters of the IMM would be sometimes in America,
sometimes in England, sometimes on the high seas and sometimes
on a golf course in Scotland. They would be, Ismay famously said,
'wherever *I* am'.

In an attempt to envisage the new president's working life, the
satirical journal *The Syren* ran, in October 1905, an imaginary interview with J. Bruce Ismay. He is pictured in a 'simple apartment,
about 150ft in length and 80ft in breadth, fitted and furnished plainly,
indeed austerely'. The walls are panelled in the 'rarest mahogany,
inlaid with mother of pearl', the light fittings are 'solid silver', and
'the furniture is priceless. Over one of the massive mantelpieces

hangs a large oil painting by a well-known Scotch academician, entitled "Bruce Vanquishing his Enemies".' Behind what looks like a
'large pianola', but turns out to be 'a wonderful entanglement of
telephone and telegraph wires, speaking tubes, etc.', Ismay can be
found keeping 'in constant and close touch with each sub-section of
each section of each division of each department of each branch of
each Line of each Trust of the great TRUST itself'.

———————

The story goes that one summer evening later in the year, Bruce and
Florence Ismay went to dinner at Downshire House in Berkeley
Square, the home of William, now Lord, Pirrie who had been managing director of Harland & Wolff since the death of Sir Edward
Harland in 1894. An Irishman determined to make his mark on
London society, Pirrie was a famously lavish host. During the
evening, the shipowner and the shipbuilder agreed, over coffee and
cigars, to build the twins *Olympic* and *Titanic*, each 15,000 tons
bigger than the *Lusitania* and the *Mauretania*.

It's an evocative image: an opulent candlelit dinner enjoyed by
two successful businessmen and their glittering wives in the Mayfair
of the gilded age. Ismay and Florence are chauffeured down to
Berkeley Square from their own magnificent house in Chesham
Street, Belgravia; after dessert the women drift off to gossip and play
cards in the fern-filled drawing room while the men, slightly drunk,
decide to build the most enormous and fabulously equipped ships
the world has ever known. The evening at Downshire House is the
perfect accompaniment to the evening at Gustavus Schwabe's
mansion forty years earlier.

Many evenings were shared between the Pirries and the Ismays
when they were all in London, but it is unlikely that the idea for the
Titanic was born during a neighbourly supper. Nor is it necessary to
bathe the birth of the ship in such a sepia glow; there was nothing
particularly inspired or outlandish in the idea of building a bigger
ship than the one before. Technology was moving at such a rate that

every ship the White Star Line commissioned from Harland & Wolff was bigger than the last; it went without saying that their next group of ships would need to be considerably larger and more dramatic if they were to grab the headlines from the two Cunarders. Nor is it likely that the idea for the *Titanic* was conceived so late in the day. The *Lusitania* and the *Mauretania* were to be launched in late 1907: why would the power-hungry J. P. Morgan, whose ambition was to eliminate competition on the North Atlantic, wait until a few months before the launch of the already famous Cunarders before reacting to their challenge?

The part played by William Pirrie in the story of the *Titanic* is, like the similarly named William Imrie, that of the forgotten partner. Except that Pirrie was more than a partner; he was the world's leading shipbuilder and regarded Harland & Wolff as his 'personal fiefdom'.[46] It was Pirrie who was behind the deal with J. P. Morgan, Pirrie who persuaded Ismay to join the conglomerate, Pirrie who pushed for Ismay as the president of the IMM, Pirrie who probably dreamed up the *Titanic*. The appearance on the scene of J. P. Morgan made Pirrie nervous about his own future: under Bruce Ismay's leadership the White Star Line could go under and Harland & Wolff would then lose their best customer. But should White Star join up with Morgan, a good deal more work might come Pirrie's way: Harland & Wolff might build ships for the entire conglomerate. So rather than standing against the threat of a syndication, after secret discussions with Morgan Pirrie decided to promote the IMM as 'something magnificent and ingenious' and he pressurised Ismay, who had initially called the whole thing 'a swindle and a humbug', into following suit.[47] The White Star Line would sooner contemplate establishing a line to the moon, Ismay had originally told reporters, than be bought by J. P. Morgan. He now quickly changed his tune. Under the terms of the arrangement between White Star and the IMM, 'all orders for new vessels and for heavy repairs, requiring to be done at a shipyard of the United Kingdom, were to be given to Harland & Wolff'.[48] 'I do not mind confessing,' Pirrie wrote to Ismay when he at last accepted the IMM presidency, 'that it was with something very like a sigh of

relief that I heard that all had been satisfactorily settled,'[49] and the
IMM was generally known as 'the Pirrie Relief-Bill'.

————

In *Titanic*, written in two weeks and published one month after the
shipwreck, the popular journalist Filson Young gave his memorable
description of her construction at Harland & Wolff.

> For months and months in that monstrous iron enclosure there was
> nothing that had the faintest likeness to a ship; only something that
> might have been the iron scaffolding for the naves of half-a-dozen
> cathedrals laid end to end. Far away, furnaces were smelting thousands
> and thousands of tons of raw material that finally came to this place
> in the form of great girders and vast lumps of metal, huge framings,
> hundreds of miles of stays and rods and straps of steel, thousands of
> plates, not one of which twenty men could lift unaided; millions of
> rivets and bolts – all the heaviest and most sinkable things in the
> world. And still nothing in the shape of a ship that could float upon
> the sea . . . The scaffolding grew higher; and as it grew the iron
> branches multiplied and grew with it, higher and higher towards the
> sky, until it seemed as if man were rearing a temple which would
> express all he knew of grandeur and sublimity . . .'[50]

Like other writers on the *Titanic,* Young describes the ship as
sinister while the excesses of her twin – whose simultaneous construction he does not mention – tend to be depicted with the humour,
respect and affection befitting an eccentric but dependable aunt.

On 31 May 1911, Morgan and Ismay were present at Harland &
Wolff for the launch of the still unfurnished *Titanic* and the
handing over of the now completed *Olympic*. On 14 June, the
Olympic set off on her maiden voyage to New York with Ismay and
Florence on board. While Florence enjoyed society, Ismay kept an
eye on the professionalism of the crew and made a note of the areas
in which the ship was deficient, details of which he sent in

Marconigrams to the White Star office in Liverpool. Her accommodation extended over five decks; she had electric elevators, a barber's shop and a dark room for photographers. She was thoroughly modern, but she was also, as a contemporary remarked, a time machine:

> You enter the reception room which is Jacobean, the restaurant is in the Louis Seize period with beautiful tapestry on the walls, the lounge is Louis Quinze with details copied from Versailles. The reading and writing rooms are of 1770, but in pure white, with an immense bow window. The smoking room is Georgian of the earlier period . . . The various apartments are decorated in almost as many styles and combinations of styles as there are rooms to be adorned . . . You may sleep in a bed depicting one ruler's fancy, breakfast under another dynasty altogether, lunch under a different flag and furniture scheme, play cards or smoke, or indulge in music under three other monarchs, have your afternoon cup of tea in a verandah which is essentially modern and cosmopolitan, and return to one of the historical periods experienced earlier in the day for your dinner . . . If a good democratic citizen of the US thinks he can enjoy his voyage better in an Empire suit of rooms – in a more comfortable bed than the emperor ever had – or a French republican likes a royal suit of one of the Louis monarchs; or an ardent German socialist suddenly evinces a desire to travel in luxury in an Imperial suit . . . whatever the taste, the steamship company will welcome and make them comfortable, as long as they pay the fare in advance.[51]

During his tours of the ship, Ismay noticed that the crew's galley was without a potato peeler, that cigarette holders were needed in the lavatories, and that the reception room, as the most popular room in the ship, would benefit from a further fifty cane chairs and ten more tables. The mattresses on the beds were too 'springy' – 'the trouble with the beds is entirely due to their being too comfortable' – and as the companionway between the lounge and the smoking room on A Deck was not used, the space might be better converted into a few extra staterooms. A further suggestion of Ismay's was to limit the service from the pantry to the saloon to one door on each side; closing off the other doors 'will enable us to put two additional tables in the Saloon, giving an increased sitting capacity of eight people'. He recorded that the voyage out took 'five days, 16 hours, 42 minutes, average speed 21.7 . . . daily runs, 428 miles, 534, 525, 548', and before returning to Southampton he arranged for the *Olympic* to be fully coaled in New York.[52]

He was clearly not a regular passenger. But nor did Ismay have the swagger and authority of a cigar-chewing shipowner. He spent his time on board the *Olympic* rearranging the deckchairs.

That same year, *Country Life* ran an illustrated feature on Dawpool which described the pile as 'a fine and acknowledged masterpiece, familiar and honoured wherever English architecture is held in esteem'. The pictures, however, seemed to be of an empty house. Instead of photographing the rooms in which the family lived, the magazine gave its readers a look at one of the vast chimneypieces and a glimpse up a panelled staircase. The author of the accompanying article suggested that the lack of any signs of life allowed 'the archi-tectural qualities of the building to stand out in strong relief unconfused by the competing charms of the beautiful furniture and pictures', but there was no disguising the fact that Dawpool was uninhabited. Following Margaret Ismay's death in 1907, the house had become a shell. It 'had more than answered its purpose', Margaret Ismay told Shaw; 'for it had interested and amused Mr Ismay every day of his life for fifteen years'. One after another their sons refused to live there. The house, which Shaw thought might be converted into a sanatorium or a smallpox hospital, lay empty until it was sold in 1927. Then the new owner had it demolished which, due to its tremendous sturdiness, took great quantities of dynamite. Dawpool was inhabited for only thirteen of its forty years, a lifespan slightly longer than that of the ships it resembled. Nothing in the Ismays' Brobdingnagian world lasted for long.

The maiden voyage of the *Titanic* was to have been 20 March, but because of a recent accident involving the *Olympic* the date was put forward to 10 April, which had originally been scheduled as the date of her second voyage. On 20 September 1911, as she was beginning her fifth Atlantic crossing under the command of Captain Smith, the *Olympic* had collided with the HMS *Hawke*, less than half her size. A large triangular hole, eight feet by fifteen feet, was punctured in *Olympic*'s side.

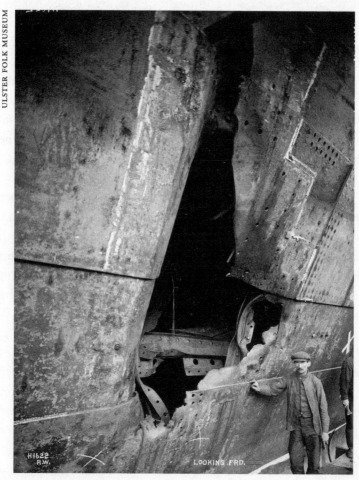

The hole in the side of the *Olympic*

Her watertight doors were ordered closed, and while two compartments were flooded the others remained dry and she stayed afloat. The passengers, who had been at lunch when their cabins were destroyed, were now offloaded and their passage cancelled. The *Olympic* was towed to Southampton where it took ten days to patch up her hull. She was then taken back to Harland & Wolff for the repairs proper to begin; because her propeller shaft, bent in the collision, needed replacing it was thought best to use the one due to be fitted to the *Titanic*. The collision cost White Star Line £130,000 in repairs and £154,000 in lost revenue when the next three voyages had to be cancelled. The *Olympic*'s crew, who were without work for a month, were given no compensation for loss of income. The subsequent inquiry concluded that Captain Smith was guilty of recklessness. The Captain was shaken – this was his first serious accident in thirty years' service – but the *Olympic* stayed afloat. These new White Star liners, Smith now believed, were indestructible.

On 21 March, Ismay's eldest daughter, Margaret, married Captain Ronald Cheape in what the papers called the 'first society wedding of the year'. It cannot have been an easy time for Ismay who, like many unhappily married men, focused his love on his favourite daughter rather than his wife. Like Ismay, Ronald Cheape was tall and handsome with a love of guns and discipline; he was also a Scot and the newlyweds were due to move to Mull after a period in India. The marriage ceremony took place at St George's, Hanover Square, and the reception was held at the Ismays' new London home at 15 Hill Street, Mayfair. Margaret, who was initially to have joined her father on the *Titanic,* was on her continental honeymoon when the Ismay family drove down to the Southampton docks in their Daimler Landaulette on 9 April.* The depleted clan stayed the night in the South Western Hotel, and, after waving Bruce off, Florence and the children went on a motoring trip to Wales. Lord Pirrie could not join

* In June 1907, the White Star Line had moved their terminal from Liverpool to Southampton to make it easier for the smart passengers from London to reach.

Ismay on the *Titanic* because he was recovering from an operation for an enlarged prostate gland; James Ismay, too, was ill (with pneumonia), and J. P. Morgan was forced to cancel at the last minute due, he said, to the pressure of work.[53] It seems that Ismay, who now took over Morgan's stateroom, was also looking for a get-out clause: two weeks before the departure date he wrote to Philip Franklin, vice-president of the IMM, suggesting that he call off his voyage, 'in view of the threatened investigation of the Shipping Companies'. Franklin replied: 'Confident no reason to alter your plans.'[54] As she left her berth in Southampton harbour, the suction from the *Titanic* snapped the mooring lines of the neighbouring ship, the *New York*. Action from Captain Smith, the pilot and a tug managed to prevent the inevitable collision.

This was to have been Ismay's final voyage in a professional capacity; seeing no future in the IMM he planned to announce his retirement on 31 December 1912. In the autumn of 1911, he offered to hand Harold Sanderson the chairmanship of the White Star Line and presidency of Morgan's combine. 'I will not attempt to disguise the fact that having been identified with the White Star Line so long and so intimately, the prospect of terminating the connection causes me real distress,' Ismay wrote, 'and I dislike to think of it; but, on the other hand the strain of the Liverpool work is, I know, beginning to tell on me . . . I hope that, upon reflection, you will not harbour the thought that I am deserting the ship prematurely.'[55] Sanderson, believing that his own career had reached a plateau, was grateful to take Ismay's place, after which Ismay changed his mind about the date of his retirement. It would now not be until 30 June 1913 because, he explained to Sanderson, 'I can only look upon my prospective severance from the business with which I have been connected all my career with very mixed and doubtful feelings, and, perhaps selfishly, I am anxious to make it as easy as possible . . . I feel that making such an entire change in my mode of life as that contemplated would come less hardly if made in the summer than in the winter, as in the former case, I should have good weather, long days, and my shooting to look forward to, which would give me occupation for some

months and this would enable me to better prepare for the time
when I should have little or nothing with which to fill up my time.'
But, Ismay conceded, 'the 30th June, 1913 is a "FAR CRY" and much
may happen between now and then'.[56] His retirement was to be kept
a secret from the IMM.

He was forty-nine and lost in the middle of his life; these are the
years in which Dante describes falling 'into a trouble that was to grip,
occupy, haunt, and all but devour me'. When Ismay boarded the
Titanic, he had betrayed his father's dream, he had discussed his
resignation with Sanderson, and he had given away in marriage the
only one of his children to whom he felt close. His own marriage was
troubled and he had set in motion a future in which he had 'little or
nothing' to look forward to but a prolonged emptiness. It is easy to
show courage if you are part of a group or representing something in
which you believe, but Ismay was alone on board and representing
nothing, no one. He had no family to wave off in a lifeboat, no son
present to whom he could set a fine example of manliness. He was
neither passenger nor owner.

When he jumped from the *Titanic*, Ismay had no status at all.

4

These Bumble-like Proceedings

Nothing more awful than to watch a man who has been found out, not in a crime but in a more than criminal weakness.

Joseph Conrad, *Lord Jim*

Enter MARINERS wet

The Tempest, I, i

The Waldorf-Astoria had originally been two hotels. The thirteen-storey Waldorf was built in 1893 by William Waldorf Astor, to irritate his aunt who lived next door. When she duly moved uptown, his cousin, John Jacob Astor – described as the world's greatest monument to unearned income – added four storeys to her house and turned it into the Astoria. In 1897, the twin buildings conjoined and, connected by an interior street known as Peacock Alley, became the largest and most luxurious hotel in the world. 'The last word in grandeur', the Waldorf-Astoria – which occupied the space of an entire city block and would later be demolished to make way for the Empire State Building – did not look like a hotel. It was described in a novel written in 1905 called *The Real New York*, as resembling 'nothing so much as a huge iceberg of gingerbread – what Lewis Carroll would have called a "gingerberg"'.[1] The interior was a stage set and professional guides were employed to give tours of the Marie-Antoinette drawing room, which mirrored the original in

Versailles, the Astor Room, which reproduced the dining room of the family home before it became a hotel, the replica Louis XV gallery and the duplication of the Soubise ballroom in Paris. 'There were more wonders,' reported the *Sun* after the hotel's official opening, 'than could be seen in a single evening – magnificent tapestries, paintings, frescoings, wood carvings, marble and onyx mosaics, quaint and rich pieces of furniture, rare and costly tablewear . . . Louis XIV could not have got the like of the first suite of apartments set apart for the most distinguished guests of the hotel. There is a canopied bed upon a dais, such as a king's bed should be . . . There are baths, elevators, electric lights.' And to enter it all, you go through revolving front doors which are like 'screw propellers'.

The purpose of the Waldorf-Astoria, as one wit put it, was to purvey 'exclusiveness to the masses'. Only incidentally somewhere to stay the night, it was a restaurant before the days when eating in public was fashionable, it was 'the club of all clubs', the place to be

Advertisement for the *Olympic* and *Titanic*, 1911

seen. The hotel welcomed unescorted women, who promenaded in pairs down Peacock Alley, or came alone simply because they could. J. P. Morgan was a patron; here the Steel Corporation was born. The hotel contained 40 reception rooms, 1,000 bedrooms and 700 bathrooms; it was thought 'big enough to hold the whole population', but it could hold less than half the number who could be held by the *Titanic*, whose first-class survivors were housed in the Waldorf on the night the *Carpathia* landed. The doorman on duty recalled how 'never before in all its history did the hotel witness such dramatic scenes as were enacted in the corridors and lobbies. So packed and jammed was the hotel that it was difficult to find room to move around.'[2] John Jacob Astor, returning with his pregnant teenage bride from their European honeymoon, was now dead and his gingerberg would provide the opening scene for the Senate's interrogation of the gilded age.

When Ismay, flanked by bodyguards along with two of the IMM's top attorneys and Philip Franklin, stepped through the Waldorf-Astoria's revolving doors at 9.30 a.m. on Friday 19 April, John Jacob Astor's final moments were already being imagined in papers across the country. In one version, he placed his wife in a lifeboat and then 'with a military salute, turned back to take his place on the sinking *Titanic*'. Another had him proclaim, after conferring with four other important men, Archie Butt, Benjamin Guggenheim, John Thayer and George Widener, 'Not a man until every woman and child is safe in the boats.' A steerage passenger described how he had been placed safely in a boat by someone he believed to have been Astor, and a song told how the millionaire parted from his wife with the words 'Good-bye my darling, don't you grieve for me/, I would give my life for the ladies to flee'. Most likely is the account given by Second Officer Lightoller of turning Astor away from the lifeboat when he asked whether he might join his wife, who was in a 'delicate condition'. 'Now,' wrote the *Denver Post*, 'when the name of Astor is mentioned, it will be the John Jacob who went down with the *Titanic* that will come first to mind; not the Astor who made the great fortune, not the Astor who added to its greatness, but John Jacob Astor, the hero.' The heroic John Jacob Astor had replaced the decadent John Jacob Astor, the man who had divorced his first wife, whose second marriage six months before to a girl younger than his own son had been considered a scandal, and whose giant hotel, frequented by fops and feminists, was considered to be a site of sexual transgression and social disorder. The manly death of John Jacob Astor had reaffirmed conservative values.

The chandeliered East Room was cleared of its gilt and brocade furniture and a walnut conference table was placed in the centre. Ismay, in a new blue suit with a black scarf running through the high collar, was the first to push through the crowds outside and his party sat and waited. Soon after 10 a.m. the doors opened to allow in reporters and a great tide of spectators, most of them well-heeled women in plumed hats. The notorious figure they had all come to see was fiddling nervously with his moustache and shirt cuffs. One journalist described Ismay as looking 'distinctly oriental', another thought he looked suspiciously 'German', and a third said his appearance was that of a 'cultivated Englishman', which was not intended as a compliment. But the general impression Ismay gave to the crowd whose gaze had fixed itself upon him was of a man who, as a fourth reporter put it, lived 'a life of ease rather than one of strength, as if he were accustomed to having his own way because it is given him rather than because he wins it'. Meanwhile in London, a crowd of 5,000 had formed around St Paul's Cathedral where a memorial service for the victims of the wreck was taking place.

At 10.30 Senators Smith and Newlands arrived with their advisor, General Uhler of the Commerce Department's steamboat inspection service. The room fell silent, and after Smith's brief introduction to the proceedings, the cross-examination of Ismay began.

In a voice scarcely above a whisper, he gave his name, his age (*fifty on the 12th of December*) and his occupation (*shipowner and*

Managing Director of the White Star Line). Asked whether he was 'officially designated to make the trial trip of the *Titanic*' Ismay replied that he had been *a voluntary passenger.* 'Will you kindly tell the committee,' asked Smith, 'the circumstances surrounding your voyage, and, as succinctly as possible, beginning with your going aboard the vessel at Liverpool, your place on the ship on the voyage, together with any circumstances you feel would be helpful to us in this inquiry?'

Ismay did not immediately answer the question. Instead, he gave a pre-rehearsed statement, doubtless drafted by Franklin:

In the first place, I would like to express my sincere grief at this deplorable catastrophe. I understand that you gentlemen have been appointed as a committee of the Senate to inquire into the circumstances. So far as we are concerned, we welcome it. We court the fullest inquiry. We have nothing to conceal; nothing to hide. The ship was built in Belfast. She was the latest thing in the art of shipbuilding; absolutely no money was spared in her construction. She was not built by contract. She was simply built on a commission.

It was the wrong time to promote the qualities of the *Titanic*, but Ismay was not used to talking in any other way. *She left Belfast, as far as I remember – I am not absolutely certain about these dates – I think it was on the 1st of April. She underwent her trials, which were entirely satisfactory. She then proceeded to Southampton; arriving there on Wednesday.* Smith, who now understood that he had made an error in not knowing from which port the *Titanic* had left England, asked Ismay to describe the trials. *I was not present,* Ismay said, and instead he gave a breakdown of the ship's progress on her first few days. *She arrived at Southampton on Wednesday, the 3rd I think, and sailed on Wednesday the 10th. She left Southampton at 12 o'clock. She arrived in Cherbourg that evening, having run over at sixty-eight revolutions. We left Cherbourg and proceeded to Queenstown. We arrived there, I think, about midday on Thursday. We ran from Cherbourg to Queenstown at seventy revolutions. After embarking the mails and passengers, we proceeded at seventy revolutions. I am not absolutely clear what the first day's run was, whether it was 465 or 484 miles. The second day the number of revolutions was increased. I think the number of revolutions on the second day was about seventy-two. I think we ran on the second day 519 miles. The third day the revolutions were increased to seventy-five, and I think we ran 546 or 549 miles. The weather during this time was absolutely fine, with the exception, I think, of about 10 minutes' fog one evening. The ship sank, I am told, at 2.20. That, sir, is all I can tell you.*

Ismay paused, and then remembered that he did have more to say. He had omitted the most important point. *I understand it has been stated that the ship was going at full speed. The ship never had been at full speed. The full speed of the ship is seventy-eight revolutions. She had not all her boilers on. None of the single-ended boilers were on. It was our intention, if we had fine weather, on Monday afternoon or Tuesday, to drive the ship at full speed. That, owing to the unfortunate catastrophe, never eventuated.* Smith did not ask about the meeting in which the decision to drive the ship at full speed was made. What he understood from Ismay's statement is that the *Titanic* was instructed to go steadily faster, and that although Ismay was only a *voluntary passenger*, it had also been his – *our* – intention to drive the ship at full speed the following week. Who, Smith wondered, controlled the ship: Mr Ismay or Captain Smith?

Asked to describe his actions after the 'impact or collision', Ismay repeated the account he had given to reporters the night before of how he had been drifting off to sleep, had got out of bed, put on his coat over his pyjamas, and gone to the bridge where he was informed by Captain Smith that they had struck ice. Nothing more was said between the two men. Ismay then walked along the starboard side of the ship, and told one of the officers to begin preparing the boats. He assisted, *the best I could* in helping to load the women and children. *I stood upon the deck practically until I left the ship in the starboard collapsible boat, which was the last boat to leave the ship, so far as I know. More than that I do not know.* A murmur of surprise went round the room; according to a number of witnesses whose comments had appeared in newspaper reports, Ismay had left in the first boat. Smith questioned him about the position of his stateroom and then returned to the purpose of his presence on the *Titanic*. Was it to view the ship in action, or did he have business in New York? *I had no business to bring me to New York at all. I simply came in the natural course of events, as one is apt to do, in the case of a new ship, to see how she works, and with the idea of seeing how we could improve on her for the next ship we are building.* The irony of his answer cannot have been missed. It appeared that Ismay was the only person on the voyage who successfully achieved what he set out to do.[3]

Had he, Smith asked, 'consulted with the Captain about the movement of the ship?' *Never*, Ismay replied. Did the Captain consult with Ismay about the movement of the ship? *Never*, he repeated, and then demurred. *Perhaps I am wrong in saying that. I should like to say this: I do not know that it was quite a matter of consulting him about it or his consulting me about it, but what we had arranged to do was that we would not attempt to arrive in New York at the lightship before 5 o'clock on Wednesday morning.* Ismay needed to put paid to the suggestion that he had been pressurising the Captain to make a record crossing by insisting that there was nothing to be gained by docking in New York any earlier than planned; it would be inconvenient to the passengers who were expecting to be picked up on Wednesday morning to arrive instead on Tuesday night and that kind of nuisance would not reflect well on the White Star Line. In which case, asked Smith, why were the ship's revolutions increased during the journey? Because, replied Ismay, *the* Titanic *being a new ship, we were gradually working her up. When you bring out a new ship you naturally do not start her running at full speed until you get everything working smoothly and satisfactorily down below.* Ismay said he had nothing to do with the 'movement of the ship', but he continued to refer to himself and the Captain as *we*; they were a team.

Did Ismay know of the ship's proximity to icebergs? *No, sir, I did not. I know ice had been reported.* Had he been on the 'so-called northern route before?' *We were on the southern route, sir.* You were not on the extreme northern route? *We were on the extreme southern* *route for westbound ships.** Made to look a fool, Smith determined to embarrass Ismay. 'What was the longitude and latitude of this ship? Do you know?' Preferring to appear ignorant, Ismay replied that he had no knowledge of such things because he was *not a sailor.* The audience duly registered that the Managing Director of the White Star Line did not have so much as a schoolboy's grasp of navigation. Did Ismay know they were near icebergs at all on Saturday? *On Saturday? No sir.* Did he know anything about a wireless message from the *Amerika* warning of ice? *No, sir.* Was he aware of their proximity to ice on Sunday? *On Sunday? I did not know on Sunday. I knew that we would be in the ice region that night sometime.* That you would be or you were? *That we would be in the region on Sunday night.* Did he have any consultation with the Captain on the matter? *Absolutely none . . . it was absolutely out of my province. I am not a navigator. I was simply a passenger on board the ship.* Did he know how the wireless service worked on the ship? *No,* he did not. Did he ever see the wireless operator? *No,* he did not. Was he on the deck when the order was given to lower the lifeboats? *Yes,* he heard the Captain give the order. Did he see any of the boats lowered? *Yes, three.* Were the boats on various decks? *No, only on the sun deck.* On the sun deck? *Yes, what we call the sun deck or the boat deck.* They were on the boat deck? That would be the upper deck of all? *The upper deck of all, yes.* Was there any order or supervision to the lowering of the boats? *Yes.* What was it? *That I could not say. The boats were simply filled and lowered away.* Did they first put in some men for the purpose of controlling them? *We put in some of the ship's people.* Some of the ship's people? *Yes.* How many? *That I could not say.* About how many? *I could not say.* How many men were in the boat in which Ismay had left the ship? *Four.* Besides yourself? *I thought you meant the crew.* I did mean the crew. *There were four of the crew.* What position did these men occupy? Ismay did not know.

* The northern track, 200 miles shorter than the southern track, was followed between August and December. The southern track was followed for the rest of the year.

How did it happen that women were first put aboard these life-boats? *The natural order*, said Ismay, *would be women and children first*. Was that the order? *Oh yes*. That was followed? *Yes*. So far as you observed? *Yes.* Were all the women and children accommodated in these lifeboats? *I could not tell you, sir*. How many passengers were in the lifeboat in which you left the ship? *About forty-five*. Forty-five? *Yes*. Is that its full capacity? *Practically, yes*. Was there any struggle around the boats? *I saw none*.

There was palpable astonishment in the room. Many of the spectators were holding their copies of the *New York Times* which that day carried an interview with Abraham Hyman, who had shared Ismay's boat. Hyman described the loading of Collapsible C as taking place amongst 'so much confusion that nobody knew what was going on . . . some of the people were too excited to understand what was said to them and they crowded forward and then some of the officers came and pushed them back, crying out for women to come first, and some of them said they would shoot any man who tried to get into the boats'.

Smith continued: you helped put some of the women in the boats yourself? *I put a great many in*. Did you see the first lifeboat lowered? *That I could not answer, sir. I saw the first lifeboat lowered on the starboard side. What was going on on the port side I have no knowledge of*. Did the first lifeboat contain the necessary number of men? *As to that I have no knowledge*. Is it true the women in the second lifeboat were obliged to row that boat from 10.30 at night until 7.30 in the morning? *The accident*, Ismay said, *did not take place until 11 o'clock*, but as to the women doing the rowing, *of that I have no knowledge*. You have no knowledge of that? *Absolutely none, sir*. So far as your observation went, would you say that was not so? *I would not say either yes or no; but I did not see it*. How long were you on the ship after the collision occurred? *That is a very difficult question to answer, sir. Practically until the time – almost until she sank*. What were the circumstances of your departure from the ship? *In what way?* Did the last boat that you went on leave the ship from some point near where you were? *I was immediately opposite the lifeboat when she left*.

Immediately opposite? *Yes.* What were the circumstances of your departure from the ship? Smith repeated, I merely ask that— *The boat was there,* Ismay interrupted. *There was a certain number of men in the boat, and the officer called out asking if there were any more women, and there was no response, and there were no passengers left on the deck.* Smith was confused; 1,500 people were left on the ship but there were no passengers left on the deck? *No sir; and as the boat was in the act of being lowered away, I got into it.* At the time that the *Titanic* was sinking? *She was sinking.* Did you see any of the men passengers on that ship with life preservers on? *Nearly all the passengers had life preservers on.* All that you saw? *All that I saw had life preservers on.* All of them that you saw? *Yes, as far as I can remember.* Naturally, you would remember that if you saw it? When you entered the lifeboat yourself, you say there were no passengers on that part of the ship? *None.* Did you see any struggle among the men to get into these boats? *No.* Was there any attempt, as the boat was being lowered past the other decks, to have you take on more passengers? *None, sir. There were no passengers there to take on.*

The East Room listened spellbound as Ismay described leaving behind him an empty ship.

Smith then asked about the damage to the *Titanic,* about the course taken by Ismay's lifeboat, about the length of time he had been on the open sea, about the number and type of boats, about how he climbed on board the *Carpathia.* He asked how many of the crew were saved, how the ship sank, whether she was broken in two, whether there was an explosion, whether there was much confusion on board as she went down, and how far he was from the *Titanic* when she took her final dive: *I did not look to see,* Ismay replied; and *I do not know how far we were away. I was sitting with my back to the ship. I was rowing all the time I was in the boat. We were pulling away.*

At this point, General Uhler looked bothered and scribbled a note which he passed to Smith. Ismay pointed out that had the *Titanic* not received her Board of Trade passenger certificate she would not be allowed to travel, adding that for all he knew she was carrying more than the recommended number of lifeboats and that lifeboats are built not for the number of passengers but to have a *certain cubic capacity.* Alexander Carlisle, who had been the Chief Designer of the *Titanic* but was now no longer employed by Harland & Wolff, would tell the British press that the final decision for the number of life-boats had been made by Ismay; that the ship's davits were able to hold forty-eight boats but that Ismay had insisted on carrying the Board of Trade requirement of sixteen along with four collapsibles.

Ismay then denied having anything to do with the selection of men who accompanied him in the lifeboat and repeated, to the astonishment of the committee, that his ship was unsinkable, that she was *specially constructed to float with any two compartments full of water. I think I am right in saying that there are very few ships – perhaps I had better not say that, but I will continue now that I have begun it – I believe there are very few ships today of which the same can be said.* While the press were noting this down, Ismay provided them with another headline: *If this ship had hit the iceberg stem on, in all human probability she would have been here today.* A gasp went round the room. This was a surprisingly informed conclusion from a man who claimed to have *no technical knowledge.* The blame for the disaster, Ismay was implying, lay in the instructions given from the bridge. Had Ismay been in the position of Chief Officer Murdoch, who was no longer living, he would not have ordered Quartermaster Hichens, the man at the wheel, to go 'hard-a-starboard'. He would have taken a risk and continued forward.

'If she hit the iceberg head on,' Smith repeated slowly, 'in all prob-ability she would be here now?'

I say, confirmed Ismay, *in all human probability she would have been afloat today.*

Smith paused and then opened and read the note passed to him by General Uhler. 'I understood you to say a little while ago that you were rowing with your back to the ship. If you were rowing and going away from the ship, you would naturally be facing the ship would you not?'

Ismay winced at the suggestion that he had taken the place of a woman and not even rowed with the men. *No. In these boats some row facing the bow of the boat and some facing the stern. I was seated with my back to the man who was steering, so that I was facing away from the ship.* Uhler, who knew this was not how you rowed, let the issue go. Smith also put the question of Ismay's conduct in the lifeboat aside for the moment but returned to it that afternoon, when he would ask Lightoller whether Ismay's claim about rowing with his back to the ship was possible. The loyal Second Officer confirmed that it was.

Ismay was now asked to describe his actions on the evening of Sunday 14 April. At what time did you dine? With whom? Did the Captain dine with you? Did you see any ice? Do you know what proportion of women and children were saved? (*I have no idea. I have not asked. Since the accident I have made very few inquiries of any sort.*) Did you interfere with the wireless communication when you were on board the *Carpathia*? (*I was never out of my room from the time I got on board the* Carpathia *until the ship docked here last night.*)

Ismay, who had presumably not slept the night before, had been on the stand for three hours. 'I thank you,' said Senator Smith, satisfied that his witness was concealing something, 'for responding so readily this morning, and for your statements; and I am going to ask you to hold yourself subject to our wishes during the balance of the day.' The newsmen rushed out to file their stories and the audience bent their heads together to confer. Ismay sat down, relieved that the ordeal was over. He would now be able to return home and put this whole business behind him.

The next day a reporter for the *New York Times* described Ismay as appearing 'cool and debonair', fielding Smith's questions with 'a smile upon his face'. The *New York Tribune* noted that he 'suffered somewhat from an unfortunate mannerism, a somewhat supercilious expression and rather too much evidence of amusement at the "land-lubberly" errors of the committee'. Another paper described him as close to 'a complete breakdown'. In Washington, Senator McCumber of North Dakota complained that 'Yesterday, one of the survivors from the lost ship was tried, convicted and executed in the Senate of the United States. I wish to register my protest against this action,

and against the condemnation or denunciation of any of the survivors or surviving officers and seamen.' Senator Newlands meanwhile
received a letter from an eminent Boston historian who insisted that
'Ismay is responsible for the lack of lifeboats, he is responsible for the
Captain who was so reckless, for the lack of discipline of the crew, for
the sailing directions given to the Captain which probably caused his
recklessness', while J. Pierpont Morgan, on holiday in France,
received a wire from his son: 'Newspapers, which are unspeakably
bad, and Congress which is worse, seem to have made up their
mind . . . Ismay is to blame for the whole thing.'[4]

With his scatter-gun style of interrogation, William Alden Smith was
authoring the official version of the story of the *Titanic*, a narrative
which would unfold over eighteen days, and fill 1,100 pages of testimony provided by eighty-three witnesses. Ismay, who had thought
– as he later said in a statement to the press – that he was needed at
the stand simply 'to ascertain the cause of the sinking of the *Titanic*
with a view to determining whether additional legislation was
required to prevent the recurrence of such a disaster', believed he had
furnished the inquiry with the information they required. But Smith
showed little interest in Ismay's authority, and the crowds who
pressed themselves into the Waldorf-Astoria had not come to discover
the cause of the wreck or to consider how to prevent such a disaster
in the future. They were here to see the state of one man's soul, to
observe a man face to face with the horror of his own limitations. 'I
did not suppose', Ismay complained, that 'the question of my
personal conduct was the subject of the inquiry.'

For Senator Smith, his opening scene could not have been more
unexpected, or more pleasing. Here we have the worst shipwreck the
peacetime world has ever known, an event discussed in every home
from Little Rock to Liverpool, and Ismay, who is lucky enough to be
alive, has nothing whatever to say about it. He speaks about what is
natural – *I came in the natural course of events; you naturally do not*

start a new ship running at full speed; the natural order is women and children first – and yet his behaviour is entirely unnatural. The instinct of other men was to protect the helpless but Ismay's was to save himself. Even the direction of his rowing is perverse; it is the wrong way round. In his speech and actions Ismay has been turned inside out. His answers all seem to be endings: *That, sir, is all I can tell you; that is all I know; I have no idea; I did not look to see; I really could not say.* He speaks only in negatives – *never, no, none, not, nothing.* He does not know the names of the officers who have died; he does not know how many women and children were left on the ship; he does not know whether the wireless operators are alive or dead – *I really have not asked;* he does not know how many lifeboats have been picked up; he did not look as his ship went down – *I did not wish to see her go down . . . I am glad I did not . . . My back was turned to her.* Commenting on his replies to questions in the witness stand, several of the next day's papers ran a list on their front page headed: 'What Bruce Ismay did not see as he left the *Titanic* with the women and children.'

> 'I saw no passengers in sight when I entered the lifeboat'
> 'I did not see what happened to the lifeboats'
> 'I did not look to see after leaving the *Titanic* whether she broke in two'
> 'After I left the bridge I did not see the Captain'
> 'I saw nothing of any explosion'
> 'I saw no trouble, no confusion'
> 'I did not recognise any of the passengers on the *Titanic* as she sank'
> 'I saw no women and children waiting when I entered the lifeboat'

The more simple Ismay's answers, the more complicated the matter seemed to be. It was as though another, octopus-like, story was wrapping its warty tentacles around his own blameless narrative and squeezing out the air. Ismay's non-answers resemble nothing so much as the trial of the Knave of Hearts in Lewis Carroll's *Alice's Adventures in Wonderland*, where the King confuses what is important with what is unimportant: '*Un*important, your Majesty means, of course,' says the White Rabbit. '"*Un*important, of course, I meant," the King hastily said, and went on to himself in an undertone, "important—unimportant—unimportant—important" as if he were trying which word sounded best.'

Ismay had nothing to conceal and everything to conceal, he was both a passenger and not a passenger; he consulted with the Captain about speed and he did not consult with the Captain about speed; the speed was increasing on Sunday and it was not increasing; he was helping to load the boats and he was not helping to load the boats; he saw the Captain but he did not see the Captain; he knew on Sunday that they were near icebergs and he did not know on Sunday that they were near icebergs; he was moving on the deck but he was motionless; there were no passengers left on board when he jumped and there were hundreds of passengers left on board. Important, unimportant, unimportant, important.

Senator Smith had stage-managed proceedings so that Captain Rostron's testimony followed that of Ismay. While Ismay had told the inquiry what he did not see, did not do, and did not know in relation to his ship, passengers and crew, Rostron described all he did to save the *Titanic*'s people; and while Ismay tried to conceal his emotional state from the Senators, Rostron openly wept. In the crisis, Ismay had panicked while Rostron thought clearly; on the stand, Ismay struggled with his meanings while Rostron thrilled the room with a plain account of how he ran his ship under a full head of steam to reach the scene of the wreck. 'Would you have done so in the night time?' asked Smith.

'It was in the night time,' Rostron politely replied. 'Although I was running a risk with my own ship and my own passengers, I also had to consider what I was going for.'

'To save the lives of others?'

'Yes,' Rostron nodded. 'I had to consider the lives of others.'

'You were prompted,' said Smith, 'by your interest in humanity. And you took the chance.'

'It was hardly a chance,' said Rostron. 'Of course it *was* a chance, but at the same time I knew quite what I was doing.'

'I think I might say, for my associates,' Smith looked around him, 'that your conduct deserves the highest praise.'

It had been 'absolutely providential', said Rostron, that the modest *Carpathia* picked up the mighty *Titanic*'s CQD (the maritime distress call). Senator Smith agreed that it had indeed been 'a very remarkable coincidence', 'so providential as to excite wonder'. Had the message come five minutes later, 'the ill-paid operator' who 'snatched this secret from the air, would have forgotten his perplexities in slumber'.

Ismay listened as Rostron (mistakenly) explained that a boat the size of Collapsible C could contain up to seventy-five people – considerably more than the forty-five he had stated earlier – and confirmed that 'if the *Titanic* had hit the iceberg bow-on – she should have been in the New York harbour instead of at the bottom of the sea'.

Smith then fired his killer question: 'Who is the master of a ship at sea?'

'By law,' replied Rostron, 'the Captain of the vessel has absolute control, but suppose we get orders from the owners of the vessel to do a certain thing and we do not carry it out: the only thing then is that we are liable to dismissal.' Smith took it in: to the crew, the Captain was next to God, to the owner, the Captain was an employee.

Rostron was released, and he returned to begin again his ship's delayed Mediterranean cruise. He was later given honorary American status when he became the first Englishman to have a plaque of his head placed in New York's Hall of Fame.

Before the morning session ended, Congressman J. A. Hughes of West Virginia made an intervention. Hughes, whose newly married daughter, Mrs Lucian P. Smith, had been on the *Titanic* and lost her husband, was a recipient of the mysterious Marconigram from the 'White Star Line' stating that the steamer was being towed to Halifax with all her passengers on board. It was Mrs Lucian P. Smith who then told the press, in a story which had spread across America, that on being picked up by the *Carpathia*, Ismay had insisted he be given a good meal and private quarters. The Congressman now read aloud a message he had received from a certain newspaper: 'You are quoted in press reports declaring, following Mrs Smith's story, that Ismay should be lynched. Please wire us, day press rate collect, 500 words, your view of *Titanic* disaster.'

Hughes wished it known that he denied using that exact wording, and had turned down the suggestion that he provide the paper with any further views. 'Lynched' was a loaded term. White supremacists were busy lynching African Americans at this time, and the most recent public lynching had taken place on the night the *Titanic* sank. It was a term which expressed visceral hatred, and which comprised the wish to see a slow execution before a bloodthirsty mob of spectators. Senator Smith thanked Congressman Hughes, and Ismay left for lunch.

———

The *Titanic* is, amongst other things, a story of doubles and so it is appropriate that the Captain of the ship should have the same name as the man who then steered the American inquiry, and that what began as a contrast between two men – the first a villain, the second a hero – should continue as a clash between two cultures, one seen as arrogant and backward-looking, the other as naïve and progressive. Ismay, who in England was not considered well bred, symbolised for Americans the moral corruption of the Old World. Senator William Alden Smith, who was regarded in America as an altruist and a seeker after truth, represented to the English the crude self-interest of the New World.

Smith's background made him the epitome of self-reliance. When he was twelve, his family, who were poor and devout, moved from the sleepy backwater of Dowagiac, Michigan, to the industrial city of Grand Rapids. Soon afterwards his father died of lung disease and William Alden dropped out of school to sell newspapers, deliver telegrams, and run a successful popcorn stall in order to support his mother and siblings. At twenty-one he started studying law (paying his way by cleaning the offices) and at twenty-four he set up his own law firm where he gained a reputation for winning his cases 'by wearing his adversaries out'. He became a Congressman and then, aged forty-seven, a maverick member of the Senate. A Republican supporter of small businesses and a champion of 'the little man', Smith fought against the likes of J. Pierpont Morgan. Nothing would give him greater satisfaction than to watch the House of Morgan sink: if the inquiry were able to prove that Ismay was negligent, or had been cognisant of negligence, on board the *Titanic* then the IMM could be sued. As a reporter for the *Grand Rapids Evening Press* explained: 'The Senator's viewpoint is that . . . the question is not one of responsibility merely, but of liability for damages in civil suits. Should it be developed that reasonable diligence was not exercised in sailing the *Titanic*, the families of survivors have a good chance to collect the damages.' Smith believed that Ismay was hiding something, and his object was to keep him in the United States until the inquiry was over, even if he was no longer required as a witness. Once Ismay returned to England he would be out of the reach of US law. As long as he remained in Washington, Smith could crack him like an egg.

The English press saw Smith as no more than a snapping terrier whose self-importance and evident ignorance could be mercilessly lampooned. 'The Michigan senator', wrote the London *Standard*, 'is less qualified as an investigator than the average individual to be picked up in the average American streetcar.' Smith was sent up as a figure in music-hall burlesque; for Joseph Conrad, who wrote about the wreck for an English literary journal, he was Mr Bumble, the cocked-hatted power-hungry beadle in *Oliver Twist*, and Conrad referred to the inquiry as 'Bumble-like proceedings'. Smith's only British defendant was Alfred Stead, whose father, the journalist W. T. Stead, had gone down with the ship. 'The newspapers,' Alfred Stead wrote, 'tell us that Senator Smith . . . is a "backwoodsman", ignorant of all nautical affairs. I do not care if he is a Red Indian. His ignorance, if it exists, is excusable ignorance, whereas the ignorance of officers and seamen in their duties is criminal negligence.'

A cartoon of Senator Smith, published in the *Graphic*, 1912

But it was not the difference between Ismay and Smith, Ismay and Rostron, or England and America which lent the inquiry its peculiar quality; it was the sameness of Ismay and Second Officer Lightoller. Generally regarded as one of the heroes of the night, Lightoller had loaded the lifeboats on the port side and then, as the ship was

descending, had taken 'a dive' and found himself drawn, by a sudden
rush of water, to the wire mesh of a giant air shaft on which he
became glued by the pressure of the sea. Unable to detach himself,
Lightoller assumed that this is how he would die when a blast of hot
air came up the shaft and blew him back to the surface of the water.
He was then pulled under again, and just as he was 'rather losing
interest in things', as he later put it, he eventually surfaced by the side
of an overturned collapsible boat. Holding onto a piece of rope, he
floated alongside it until one of the ship's giant funnels fell, missing
him by inches and causing the raft, and Lightoller, to be flung fifty
yards clear of the sinking *Titanic*. Men were now starting to scramble
onto the lifeboat and Lightoller joined them, eventually taking
control. 'If ever human endurance was taxed to the limit,' he said in
his memoirs, 'surely it was during those long hours of exposure in a
temperature below freezing, standing motionless in our wet clothes.'
He ordered every man on the upturned boat to face the same way
and to 'lean to the left or stand upright or lean to the right, as the
case might be. In this way we managed to maintain our foothold on
the slippery planks by now well under water.' Here the party remained
for several hours until they were taken on board two of the half-
empty lifeboats.

Senator Smith was unmoved by accounts of Lightoller's survival;
as far as he was concerned Rostron was the saint of the story and
Lightoller simply a stooge of Ismay, more concerned with keeping
his job with the White Star Line than preventing future tragedies at
sea. Had they not been whispering together in the cabin of the
Carpathia, cooking up their plot to abscond on the next available
White Star liner without so much as setting foot on American soil?

Lightoller was called to the stand after lunch. Described by the
papers as 'strong and powerfully built' with a 'virile sea-worn face', he
told the inquiry in a clear, deep voice how he had retired to bed and
was dropping off to sleep when he heard 'a slight shock and a grind-
ing sound. That was all there was to it. There was no listing, no
plunging, diving, or anything else.' He then left his room in his
pyjamas and went to the bridge where he found the Captain and

First Officer Murdoch motionless, looking ahead. The ship was still moving and so he assumed all was well. Lightoller then returned to his room. 'What for?' asked Smith.

'There was no call for me to be on deck,' replied Lightoller.

'No call, or no cause?' corrected Smith.

'As far as I could see,' said Lightoller, 'neither call nor cause . . . I did not think it was a serious accident.' As he walked back, Lightoller saw no one except the Third Officer, 'who left his berth shortly after I did'. The two men briefly conferred about the incident and concluded that 'nothing much' had happened.

'Did you go back to your room under the impression that the boat had not been injured?' asked Smith.

'Yes, sir,' said Lightoller.

'Didn't you', wondered Smith, 'tell Mr Ismay that?'

Lightoller answered that he had not yet seen Ismay; that he would only see Ismay once, about twenty minutes later, and that he 'really could not say' whether he had or had not then told him there was no cause for concern. Lightoller went back to bed for 'ten minutes'. When he was roused by Fourth Officer Boxhall he put some clothes on top of his pyjamas and went out on deck. Ismay, when Lightoller saw him, was standing stock-still and alone. According to Lightoller the entire evening had been conducted in silence; he appears in his account like a man in a dream, sleepwalking to the bridge and then back to bed where he wakes with a start, realising that everyone is going to die.

Nor did Lightoller recall seeing any passengers on the deck when the ship was going down. 'I ask you again,' Smith persisted: 'There must have been a great number of passengers and crew still on the boat, the part of the boat that was not submerged, probably on the high point, so far as possible. Were they huddled together?'

'They did not seem to be,' said Lightoller. 'I could not say, sir; I did not notice; there were a great many of them, I know, but as to what condition they were in, huddled or not, I do not know.' However, in his memoirs, written twenty-three years later, Lightoller remembered the crowds and what condition they were in. They were washed back in a 'dreadful huddled mass . . . It came home to me very clearly how fatal it would be to get amongst those hundreds and hundreds of people who would shortly be struggling for their lives in that deadly cold water.'

Lightoller explained how the Captain gave him no order to load the lifeboats on the port side, how he had placed twenty-five people into the first boat because he was unconfident about filling it with more, how he had personally tested all the lifeboats when the ship was in Southampton (in his memoirs, he admitted to testing only 'some of them'), and how it was only as she began to list that he realised there was a genuine emergency. There had been no confusion amongst the passengers, who 'could not have stood quieter had they been in church', and no restraint on the movement of those in steerage. No other man had tried to join them on the upturned lifeboat because everyone else was 'some distance away'.

As for the question of whether he had seen any ice warnings, Lightoller did not know they were in 'the vicinity of icebergs', he 'could not say' whether he saw the 'individual message' sent by the *Amerika* on Sunday evening or heard of it even though it was received during his shift. (This particular Marconigram, which never reached the bridge of the *Titanic* from the communications room, had been intercepted and sent to Washington where it ended up on Senator's Smith's desk.) Would it not, asked Smith, 'have been the duty of the person receiving this message to communicate it to you, for you were in charge of the ship?' Lightoller replied only 'under the commander's orders, sir' and that while he did not know about that particular message he 'knew that a communication had come from some ship; I can not say that it was the *Amerika* . . . speaking of the icebergs and naming their longitude . . . the message contained information that there was ice from 49 to 51'.

'How do you know it came?' asked Smith.

'Because I saw it,' replied Lightoller. At one o'clock on Sunday, the Captain had shown it to him.

Lightoller's cocky responses during the five hours he was questioned were designed to ridicule Smith's pretence at authority. He thought the inquiry, as he put it in his memoirs, a 'complete farce wherein all the traditions and customs of the sea were continuously and persistently flouted'. It was an outrage that professional seamen should answer, before their clothes had even dried, to what he called 'an armchair judge' who had never himself 'been called upon to make a life-and-death decision in a sudden emergency'.[5]

What mattered to Lightoller was the forthcoming British inquiry; this American affair was an amateur theatrical which had to be endured. Asked at what time he left the ship, Lightoller replied that he didn't leave it, that the ship had left him. He kept a straight face when, in an account of how the ship's forward funnel had fallen on top of a group of struggling swimmers, Smith inquired whether any of them were injured (one of the passengers probably killed by the funnel was John Jacob Astor, whose body, when it was found by the recovery ship, *Mackay-Bennett*, floating in the Atlantic, was so crushed and charred that it could be identified only by the $2,440 in notes in his pocket and the diamond ring on his finger).

In a remark which was then hard for him to live down, Smith suggested to Lightoller that some of the passengers might, as a 'last resort', have tried to keep dry in the *Titanic*'s watertight compartments. 'Is that at all likely?' Smith asked.

'No sir,' said Lightoller. 'Very unlikely.'

William Alden, who had by no means finished with Lightoller, was afterwards known by the British press as Watertight Smith.

After Lightoller's testimony, Ismay left the room and paced the corridors of the Waldorf-Astoria smoking cigarettes. He was joined by journalists who badgered him as to whether he had left a ship filled with women and children. The inquiry, Ismay responded in a rage of indignation, was 'horribly, horribly unfair. I cannot understand it.' He spoke to the press – whom he knew to be conducting a trial of their own – as he had been unable to speak to Smith. Ismay, who had never before made a speech, was now fighting for his life:

What sort of man do you think I am? Do you believe I'm the sort who would have left the ship as long as there were any women and children aboard her? I think it was the last boat that was lowered I went into. I did then what any other passenger would do. And tell me how I was different from any other passengers? I was not running the ship. If they say I was the president of the company that owns the ship then I want to know where you will draw the line. As I lay in my stateroom on board the *Carpathia* I went over every detail of the affair. I have searched my mind with deepest care. I have thought over each single incident that I could recall of the wreck. I am sure that nothing wrong was done – that I did nothing I should not have done. My conscience is clear, and I have not been a lenient judge of my own acts. I took the chance when it came to me . . . Why, I cannot even protect myself by having my counsel ask questions. Don't misunderstand me by thinking I mean questions calculated to twist the witnesses up. On the contrary, I mean questions intended to simplify involved meanings. A glaring example of this happened this morning when I was asked about rowing the boat and I said I had been at one of the oars, but had not seen the ship go down. At once I was asked how I could have failed to see her since, if rowing, I must have been facing her. It would have been easy for my counsel to show that I was pushing at an oar which was being handled by two or three of us. There was no room to sit down in the proper oarsman's position.[6]

Various versions of these words appeared in newspapers around the world, and Ismay would never again speak to a journalist in so unguarded a fashion. It was unlike him to be this voluble; he had acted out of character. He had wanted, as he put it, to 'simplify' his own 'involved meanings'. He needed to talk, to unravel the knots which were tying inside him; he needed to be understood.

The press were relying on the fact that those survivors who were not called as witnesses, or those who had been called and consequently felt misrepresented by Senator Smith, were keen to tell their story; the construction of a stable narrative was a way of dealing with

the chaos of the experience. So while the inquiry was making crimi-
nals of some survivors, the press were turning others into heroes.
That evening's papers carried an interview with one of the *Titanic*'s
valiant stokers, who said that they had been told to 'fire her up as
hard as we possibly could. From the time we left Queenstown until
the moment of the shock we never ceased to make from seventy-four
to seventy-seven revolutions. It never went below seventy-four and
during that whole Sunday we had been keeping up to seventy-seven.'
In that case, Senator Smith understood, the ship had exceeded the
seventy-five revolutions maintained by Ismay.

Later that night, Ismay asked permission to return home the
following morning on the *Lapland*. Smith refused, placing him and
some thirty-five members of the crew under surveillance and serving
them with subpoenas to appear before the committee the following
week. When the *Lapland* then sailed carrying five members of the
Titanic's crew, including the steersman Robert Hichens, Smith had
the US Navy informed, the ship stopped and the witnesses were
returned to shore by pilot boat. Why, Smith wondered, was the
White Star Line so determined to get its men out of the country?

———

On the morning of Saturday 20 April, the inquiry reconvened in the
hotel's ballroom, described as the Waldorf-Astoria's 'outstanding
feature' and 'the most sumptuous apartment of its sort in New York,
if not in America'. Usually reserved for banquets, concerts and
parties, the ballroom could also be, as a contemporary writer put it,
'transformed almost instantly from a huge *palace-de-danse* into a
most comfortable and practicable theatre, with more than 1,100 little
gilt chairs, in addition to the permanent double tier of boxes around
three sides of the room'.[7] The focus of the press that morning was the
arrival of the glamorous suffragette Inez Milholland. A New York
celebrity, she was accompanying her former fiancé Guglielmo
Marconi, who was due to give evidence about the use of wireless on
the *Titanic* and the *Carpathia*.

While the behaviour of the men on the *Titanic* represented to the
popular imagination the 'natural' order confirmed by the sea, Inez
Milholland was a symbol of the increasingly unnatural order of
things on land. Aged eighteen, she had made four militant suffrage
speeches on a soap box in Hyde Park and paraded the streets of
London with a banner emblazoned with 'Votes for Women'. In 1911,
Milholland appeared in barely disguised form as the passionate
heroine of Isaac Stevenson's novel, *An American Suffragette*. Her presence at the inquiry today was a reminder of the 'Votes or Boats'
debate which had been ignited by the *Titanic* disaster: women in the
lifeboats had refused to return to rescue the men whose gallantry
they had been only too pleased to accept on the sinking ship. 'What
do women want?', the newspapers asked. It seemed that chivalry at
sea was considered chauvinism on land. 'I suggest, henceforth,' said
a man from St Louis, 'when a woman talks women's rights, she be
answered with the word *Titanic*, nothing more – just *Titanic*.' 'The
heroism of the men on the *Titanic*,' wrote the Baltimore *Sun*, shows
'that women can appeal to a higher law than that of the ballot for
justice, consideration and protection.' A writer calling himself 'Mere
Man' asked if 'the suffragette would have stood on that deck for
woman's rights or for woman's privileges?'[8] Miss Milholland had
come as the friend of Marconi, but she also wanted to hear for herself
what an anti-suffrage journal was triumphantly calling 'the story that
came up from the sea'.

First to the stand was the *Carpathia*'s wireless officer Harold
Cottam, who was questioned about the origins of the Marconigrams,
one of which had been received by Congressman Hughes, stating
that the *Titanic* was being towed to Halifax with all her passengers
safe on board. If these messages were not sent by Ismay in order to
give the White Star Line time to reinsure the ship then who, the
inquiry wondered amongst themselves, was responsible for them,
and what purpose did they serve? Cottam claimed ignorance, after
which the spectators cleared the way for twenty-two-year-old Harold
Bride, the *Titanic*'s wireless operator, who had already sold his story
to the *New York Times* for a considerable sum. Bride entered in a

wheelchair where he sat slumped, his crushed and frostbitten foot in plaster and his pale face leaning against a pillow. Ismay, seated between Franklin and Lightoller, listened to the testimony of one of the men whose survival he had not bothered to inquire about. He heard how Bride and his fellow Marconi operator, Jack Phillips, had 'joked' as they sent out CQD signals saying they were sinking by the head. 'The humour of the situation appealed to me, and I cut in with a little remark that made us all laugh, including the Captain. "Send SOS," I said, "it's the new call, and it may be your last chance to send it." We said lots of funny things to each other in the next ten minutes.'

Soon after the *Titanic* made contact with the *Carpathia,* her wireless began to grow weaker so 'I went out on deck and looked around. The water was pretty close up to the boat deck. There was a great scramble aft, and how poor Phillips worked through it I don't know. He was a brave man. I learned to love him that night, and I suddenly felt for him a great reverence to see him standing there sticking to his work while everybody else was raging about. I will never live to forget the work Phillips did for the last awful fifteen minutes. He clung on, sending and sending. He clung on for about fifteen minutes after the Captain released him. The water was then coming into our cabin.' As the band started to play 'Autumn', Phillips 'ran aft' and as Harold Bride was helping a group of the crew to lower the final collapsible boat, a wave washed them all into the sea.

Bride's testimony was interrupted at one point when the doors burst open and a woman ran crying into the ballroom. Was First Officer Murdoch alive or dead? Smith asked whether Lightoller 'would be good enough to tell the lady whatever she wishes to know'. Lightoller, who had been the last person to see Murdoch alive took her aside and explained that 'Mr Murdoch died like a man doing his duty', after which she left quietly.

Bride continued. 'The next I knew I was in the boat. But that wasn't all; I was in the boat, and the boat was upside down, and I was under it. I just remember realising I was wet through, and that whatever happened I must breathe.' He got out – 'how, I don't know' – and started swimming from the ship. 'She was a beautiful sight

then. Smoke and sparks rushing from her funnels.' The band was still
playing as the ship 'flowed' down. 'I felt after a little while like
sinking. I was very cold. I saw a boat of some kind near me, and put
all my strength into an effort to swim to it. It was hard work and I
was all alone when a hand reached out . . . and pulled me aboard.'
The hand came from the upturned collapsible lifeboat which was
being manned by Lightoller, and Bride was the last man 'invited' on
board. *Invited?* Smith expressed surprise: 'There were others struggling to get on?'

'Yes,' replied Bride who had not heard Lightoller's testimony the
day before, in which he said the opposite: 'dozens . . . men were
swimming and sinking everywhere'.

'Dozens in the water?' Smith repeated. Who were, Smith wondered,
the men on the boat?

'They were all part of the boat's crew,' said Bride. A horrified gasp
went round the room as the reality of the scene became apparent.

Ismay, sitting at one end of the table, did what he would do
throughout the hearings: on sheets of paper taken from the press
table, he drew over and over again an image of the White Star flag.

There was no session that afternoon. Instead the committee and their
subpoenaed witnesses prepared to leave for Washington, where the
proceedings would continue on Monday morning in the less lush
atmosphere of the new Senatorial buildings. That night Ismay again
asked for, and was again refused, permission to return home, at
which point he lodged a formal complaint with James Bryce, the
British Ambassador. 'I have the utmost respect for the Senate of the
United States,' Ismay said, 'but the inquiry as it is proceeding now
may wreak great injustice rather than clear up points in question.'

Senator Smith, Bryce explained to the British Foreign Office, was
responding to the parallel inquiry into Ismay's actions being
conducted by the American press, particularly the sensationalist
'Yellow Press' dominated by William Randolph Hearst. One of the

most powerful formers of opinion in America, Hearst had met Ismay when he was a White Star agent in New York. His immediate dislike of the aloof Englishman was intensified when Morgan made Ismay president of the IMM. A reformist who saw himself as a spokesman for the American people, Hearst owned papers in every region and in six major cities, all of which he treated as a daily letter to his 4 million readers. He called his work 'government by newspaper', and whatever causes he championed or campaigned against tended to win the day (the views expressed in Hearst's New York *Journal* were credited with provoking the Spanish-American War). His style was defined by one of his editors as 'Gee-Whizz Journalism': 'Any issue the front page of which failed to elicit a "Gee Whizz!" from the readers was a failure, whereas the second page ought to bring forth a "Holy Moses!" and the third an astounding "God Almighty!"'[9] Described by Theodore Roosevelt as an 'unspeakable blackguard' who combined 'with exquisite nicety all the worst faults of the conscienceless, corrupt, and monied man', Hearst ran a campaign against Ismay which presented him in these same terms. One of Hearst's publications, the *New York American*, carried a front-page cartoon of the White Star chairman sitting in a lifeboat watching a sinking ship, accompanied by the words: 'THIS IS J. *BRUTE* ISMAY: A picture that will live in the public memory for ever. J. Bruce Ismay safe in a lifeboat while 1,500 people drown. We respectfully suggest that the emblem of the White Star be changed to that of a "Yellow liver".' 'He must be glad he is an Englishman,' another of his papers said. 'He is no gladder than we are.'

'However much may have been written about this school of journalism,' Bryce explained to the British government, 'it is difficult for any person not residing in this country to realise to what extent their influence over the less well-educated classes is pernicious in its heartless exploitation of every calamity that has ever saddened, and of every scandal that has ever attracted attention in, this country.'[10] Both William Randolph Hearst and William Alden Smith would stop at nothing, Bryce believed, in their exploitation of the grief and rage of the American people so long as it served their own ends, and for each man the story of Ismay's survival was bigger than the story of the *Titanic* itself. Added to which, Senator Smith felt no need to inform Bryce of his actions or intentions. 'It might have been more courteous,' the British Ambassador felt, 'if some communication had been addressed to this embassy, though from persons so ignorant of the usages of international relations as most members of the Senate are no such action need have been expected.' And as to his qualifications for chairing the committee, it 'was generally admitted that Senator Smith is one of the most unsuitable persons who could have been charged with an investigation of this nature'. Smith was the type of man who was 'always anxious to put himself forward where any passing notoriety can be achieved'. The US Secretary of State agreed; the inquiry was 'mere self-advertisement'. Nor did Taft dispute this assessment of Senator Smith's character. The President of the United States, according to Bryce, was 'of the opinion that as long as the chairman of the committee thought it would keep him in the headlines the inquiry would go on'.[11]

As William Alden Smith left the last session of the New York hearings on the afternoon of Saturday 20 April, he told journalists on the steps of the Waldorf-Astoria that 'the surface has barely been scratched. The real investigation is yet to come.'

Ismay and his bodyguards arrived in Washington on Sunday at 8 p.m., exactly one week after the *Titanic* hit the iceberg. Hundreds of people were gathered around the entrance to Willard's Hotel in the hope of getting a look at the infamous coward, and so Ismay slipped in through the back door and was whisked up the stairs to the suite which had been reserved for his party. Lightoller was booked into the less grand Continental together with the *Titanic*'s frustrated crew, who, he said in his memoirs, 'refused point blank to have anything more to do with either the inquiry or the people, whose only achievement was to make [them] look ridiculous'. With the help of Bryce and Franklin, Lightoller managed to 'bring peace to the camp', but he was annoyed at being quartered with his inferiors rather than with the White Star officials he was doing so much to protect, and insisted on being either moved to the Willard or provided with a separate floor and dining arrangements. The Continental's staff muttered about how the rich and the poor on the *Titanic* were now bedding down side by side on the ocean floor.

The papers that weekend had run Emily Ryerson's sensational account of her encounter with Ismay when she had been walking with Mrs Thayer on the deck of the *Titanic*. 'Sunday morning Mr Ismay showed to a woman passenger a wireless message with an ice warning. She asked if the *Titanic* would not go slower and Mr Ismay replied laughing, "No FASTER! We want to get by it".' Here was the evidence the inquiry needed to prove that Ismay was party to negligence; it also implied that he had lied when he told Senator Smith that he was not aware on that day of any proximity to ice. Mrs Ryerson agreed to swear to her story, and that night Ismay issued a press release. Unlike the few words he gave to reporters when he left the *Carpathia*, it seems likely that Ismay himself drafted this next statement which appeared in *The Times* on 23 April, and that it is therefore his only published self-defence. It was 'impossible', Ismay said, 'to answer every false statement, rumour or invention that has appeared in the newspapers . . . but I do not think that courtesy requires me to be silent in the face of the untrue statements'. He stressed his initial willingness to appear at the inquiry, 'without subpoena', and to answer every question put to him 'to the best of my ability, with complete frankness and without reserve'. He described again, this time for the benefit of the British public, his actions on the *Titanic*, repeating that he had been 'a passenger and [therefore] exercised no greater right or privileges than any other passenger. I was not consulted by the commander about the ship, her course, speed, navigation, or her conduct at sea.' He stated that he had not, as claimed in the press, been at a dinner party with the Captain on Sunday night, and that it was 'absolutely and unqualifiably false that I ever said that I wished that the *Titanic* should make a speed record or should increase her daily runs'. He confirmed that an ice warning from the *Baltic* had indeed been 'handed to me by Captain Smith, without any remarks, as he was passing me on the afternoon of Sunday, April 14. I read the telegram casually and put it in my pocket,' and insisted that he had said nothing to Mrs Ryerson about the ship being speeded up. Had the warning 'aroused any apprehension in my mind – which it did not – I should not have ventured to make any suggestion to a commander of Captain Smith's experience and responsibility, for the navigation of the ship rested solely with him'. Ismay described, yet again, how he had been asleep at the time of the collision, how he worked at loading the starboard boats, and how, as the forward collapsible lifeboat was being lowered, 'Mr Carter, a passenger, and myself got in.' He explained that the Yamsi messages sent from the *Carpathia* 'have been completely misunderstood' and that he could not be accused of trying to avoid the Senate Committee's inquiry because he had 'not the slightest idea that any inquiry was contemplated'. He concluded by saying that 'it was the hope of my associates and myself that we had built a vessel which could not be destroyed by the perils of the sea or the dangers of navigation. The event has proved the futility of that hope.'

Futility had been the title of a little-read novella by Morgan Robertson, published fourteen years before. It told the story of a 45,000-ton 'floating city' called the *Titan*, the largest ship ever built, which left New York in April with 3,000 passengers. Speeding along at 25 knots she met an iceberg in the place where the *Titanic* would later meet her fate, and because the *Titan* was believed to be 'unsinkable', and carried 'as few boats as would satisfy the laws', she went down with most of her passengers still on board.[12]

The following day the inquiry reconvened in the sumptuous new Caucus Room of the Senate Office Building, where the McCarthy and Watergate hearings would in time be heard. 'Washington this time of year', reported the *Daily Telegraph*, 'is crowded with honeymoon couples, and the weather is so mild that the windows of the Senate are open all day long. The scent of spring flowers wafted into the court, and the words of witnesses are punctuated by the sweet notes of birds singing in the leafy trees outside. Ladies in spring dresses sit in court fanning themselves, and at times so crowded is the courtroom that the atmosphere is insufferably close.'

First to the stand was Philip Franklin, the forty-one-year-old American vice-president of the IMM. The wreck of the *Titanic*, Franklin said, 'has demonstrated an entirely new proposition that has to be dealt with – something that nobody had ever thought of before. These steamers were considered tremendous lifeboats in themselves.' It was, Franklin now believed, 'impossible to build a non-sinkable ship'. Smith asked Franklin to read aloud every Marconigram he had sent and received in relation to the *Titanic*, including, to Ismay's cringing embarrassment, the Yamsi messages. Their contents were duly noted down by the newsmen and reported the next day in papers around the world. The reproduction of the Marconigrams gave the story a sense of synchronicity; readers felt as though they were experiencing the disaster and its aftermath in real time. It was vital, Franklin stressed at length, that Ismay's messages not be misunderstood. It was not his own welfare he was concerned with when he wanted to return on the *Cedric*, but that of the crew. 'Criticism has been seriously made to the effect that those messages were sent entirely with the idea of getting the crew away, and of Mr Ismay's also getting away on account of what information might come out from the crew. I want to say that that was not in Mr Ismay's mind. Everybody realises the importance of getting these members of the crew away from the country at the earliest possible moment.' Franklin spoke about the danger of letting a ship's crew loose in a city, after which Smith released him. Next to the stand was Joseph Boxhall, the *Titanic*'s Fourth Officer, who recalled how, contrary to Lightoller's testimony, only two boats had been lowered during the boat drill in Southampton.

The *New York Times* that day carried an interview with Inman Sealby, formally a captain of the White Star Line and now a student of admiralty law at the University of Michigan. Sealby had been dismissed by the company in 1909 when the *Republic* sank under his command. In the twenty-five years he had worked for the Ismay family, Sealby said, 'I never saw anything which would have led me for one minute to anticipate anything but the best of conduct in Mr Ismay at all times and under all circumstances. Until the exact circumstances of his escape from the *Titanic* are placed before me it would be impossible for me to pass an opinion in the matter.'

That evening Ismay again requested permission to return home, promising that he would be replaced by an army of experts from Great Britain, including personnel from White Star and engineers from Harland & Wolff. Smith again refused. Ismay seemed to be pulling in opposing directions: on the one hand he stressed his willingness to help Smith, and on the other hand he talked of little else but leaving for England.

The acoustics in the Caucus Room had not been good enough to carry the often quiet voices of the survivors, and so on Tuesday 23 April, the inquiry moved once more, this time to the stern Committee on Territories conference room. During the last four days the proceedings had occupied four different premises, from a gilded hotel to a government office. Senator Smith's settings were impersonating the drift of the age itself, from frivolity to seriousness. The new room was too small to contain spectators, but around 500 women who had been waiting since 9 a.m. nonetheless poured through the door and were forcibly removed by police officers. This process delayed the hearings by an hour, which allowed time for photographers to get some pictures of Ismay, before being hurled out by Smith.

Third Officer Pitman described how 'crowds' of 'moaning' victims froze to death in the water while those in the lifeboats 'just simply lay there doing nothing', and Frederick Fleet, the lookout who reported the iceberg, explained how he 'had no idea of distances or spaces' and that if the *Titanic* officers had not mislaid the binoculars he would have seen the iceberg in time. Comic relief came in the testimony of a Canadian first-class passenger, Major Peuchen, who, in the absence of a member of the crew, had volunteered to man a boat. Asked whether the ship went down by the 'bow or the head', Peuchen asked: 'You mean "the head" by the bow, do you not?'

'Exactly,' said Smith.

'It is the same thing,' remarked Peuchen.

'No,' said Smith grandly. 'Not the same thing.'

Ismay now suggested to Smith that if he was not currently needed to give evidence nor allowed to return to his wife and family, might he go to New York for a few days and get on with some work? While his request was not strictly denied, the committee 'took no action'. Late that night, the *Telegraph*'s man in Washington saw Ismay in Willard's Hotel accompanied by Franklin, three lawyers and the officials of the IMM. 'Despite his terrific experiences,' the correspondent reported home, 'including the scandalous and persistent onslaughts made on him through the editorials and cartoons of the Yellow Press, Ismay bears himself bravely. His eyelids looked red and swollen, as if through lack of sleep, but otherwise the chairman of the White Star Line was just as alert, erect and dapper as when I saw him last in England.'

———

At nine o'clock on the morning of Wednesday 24 April, while the throngs of onlookers gathered like pigeons around the Congressional buildings and cinematograph wheels prepared to whirl, Ismay knocked on Smith's door and demanded to be put on the stand at once so that he could then sail to England. Smith impatiently dismissed him and opened the proceedings an hour and a half later with an announcement: 'From the beginning until now there has been a voluntary, gratuitous, meddlesome attempt upon the part of certain persons to influence the course of the committee and to shape its procedures. The committee *will not tolerate* any further attempt on the part of anyone to shape its course. We will proceed *in our own way*, completing the official record.' The newsmen pricked up their ears: who could the Senator be referring to? Meanwhile, the White Star offices in England and New York were receiving reports that the crew of the *Olympic*, currently in Southampton and waiting to sail on the same route as the *Titanic*, had mutinied. Immediately on landing, Ismay, advised by Franklin, had ordered the IMM to provide their ships with lifeboats to accommodate all passengers and crew; but where were these hundreds of surplus boats to come from? In the few days between his command and the *Olympic*'s sailing, the White Star Line had only been able to get hold of what the *Olympic*'s crew described as forty 'rotten and unseaworthy' collapsibles.

First to appear that day was twenty-nine-year-old Fifth Officer, Harold Lowe. Smith invited him to tell the story of his life, beginning with the time, aged fourteen, when he ran away to sea. 'It is a pretty long story,' Lowe laughed, and the audience craned their necks. 'Here at last,' reported the *Telegraph*, 'came a unique opportunity of hearing at first hand what promised to be a thrilling tale of romance and reality of sea life. Mr Lowe was, however, disappointing.' He described how he had spent seven years on schooners, then switched to sail, and finally earned his certificates on steamers. Lowe 'missed the chance of a lifetime', said the *Telegraph*, in reducing a 'thrilling tale' to a few facts, but the *Titanic*'s Fifth Officer was one of the few British men who did emerge from his testimony a hero. He was celebrated in England for providing the best moment of the week so far: when Smith asked him of what an iceberg was composed, Lowe replied that it was composed of 'ice'. In America he was celebrated for 'cursing out' Ismay, who had been 'interfering' with his duties. 'Lower away! Lower away! Lower away! Lower away!' Ismay shouted as Lowe was releasing one of the lifeboats on the starboard side. Lowe refused to repeat what he said in response to Ismay, whom he had not recognised. Ismay himself now interjected, and turned to Lowe. 'Give us what you said.' He then turned to Smith, 'I have no objection to his giving it. It was not very parliamentary.' If the Senator was concerned that the language might be offensive, Ismay suggested that 'it be put on a piece of paper' so that he could decide for himself. Lowe wrote it down and Smith read aloud: 'If you get the hell out of that I might be able to do something.' How did Ismay respond to Lowe's insubordination, Smith wondered? He just walked to the next boat, said Lowe, 'on his own hook, getting things ready there, to the best of his ability'. Concerned that he might have blown his future with the White Star Line, Lowe conceded that Ismay was clearly 'anxious to get the people away and also to help me'. The following day various US papers suggested that towns called Ismay in Texas and Montana should change their names, perhaps to Astor but preferably to Lowe, in honour of the officer who had told Ismay to 'go to hell'.

Smith now recalled Lightoller, who made a statement taking full responsibility for the 'Yamsi' messages sent on the *Carpathia*. Lightoller described Ismay as having been entirely passive in the doctor's cabin, as deferring to the authority of the Second Officer: 'On having a conversation with Mr Ismay he also mentioned about the *Cedric* and asked me my opinion about it, and I frankly stated that it was the best thing in the world to do if we could catch the *Cedric*. Later on he remarked that owing to weather conditions it was very doubtful if we would catch the *Cedric*. I said, "Yes, it is doubtful. It will be a great pity if she sails without us." "Do you think it will be advisable to hold her up?" I said, "Most undoubtedly; the best thing in the world to hold her up."' Lightoller then described how 'a telegram was dispatched asking them to hold the *Cedric* until we got in, to which we received the reply that it was not advisable to hold the *Cedric*'. Ismay 'asked what I thought about it. I said, "I think we ought to hold her, and you ought to telegraph and insist on their holding her and preventing the crew getting around in New York." We discussed the pros and cons and deemed it advisable to keep the crew together as much as we could, so we could get home, and we might then be able to choose our important witnesses and let the remainder go to sea and earn money for themselves. So I believe the other telegram was sent.'

'I may say,' Lightoller continued, 'that at that time Mr Ismay did not seem to me to be in a mental condition to finally decide anything.

I tried my utmost to rouse Mr Ismay, for he was obsessed with the idea, and kept repeating, that he ought to have gone down with the ship because he found that women had gone down. I told him there was no such reason; I told him a very great deal; I tried to get that idea out of his head, but he was taken with it; and I know the doctor tried, too; but we had difficulty in arousing Mr Ismay, purely owing to that wholly and solely, that women had gone down in the boat and he had not. You can call the doctor of the *Carpathia* and he will verify that statement.' It was 'naturally human nature', Lightoller said, to try 'to get the men back to their wives and families as soon as possible. Their income stops, you know, from the time the wreck occurs, legally.'

But why, wondered Smith, had Lightoller not made this statement earlier?

'Because,' Lightoller replied, 'the controversy in regard to the telegram had not been brought up then, or brought to my knowledge; I mean all this [news]paper talk there has been about this telegram.'

And that is the reason, suggested Smith, 'that you were prompted to make this disclosure?'

Lightoller replied that he was making the disclosure because he was 'principally responsible for the telegram being sent'.

'And you sent it?' inquired Smith.

'I did not,' replied Lightoller.

'You delivered it to the wireless?'

'I did not.'

'Who did?'

'I do not know.'

'Did you write it out?'

'I did not.'

'Did you speak to the operator about it?'

'I did not.'

'Have you spoken to him about it since?'

'I have not.'

'But you wish to be understood as saying that you urged Mr Ismay to send it?'

'I did.'

'Did you know at that time,' asked Smith, 'that an inquiry had been ordered by the Senate?'

'Certainly not,' said an apparently appalled Lightoller, 'or we should never have dreamed of sending the telegram. Our whole and sole idea was to keep the crew together for the inquiry, presumably at home. We naturally did not want any witnesses to get astray.' In his memoirs, however, Lightoller later wrote: 'Everyone's hope, so far as the crew were concerned, was that we might arrive in New York in time to catch the *Cedric* back to Liverpool and so escape the inquisition that would otherwise be awaiting us. Our luck was distinctly out. We were served with warrants, immediately on arrival.'[13]

Smith then returned to the question of whether the various ice reports had been taken notice of and Lightoller was as uncommunicative on the subject as he had been five days earlier. 'Did you see', asked Smith, 'in the chart room of the *Titanic* any memoranda in the rack advising that you were in the vicinity of ice?' Lightoller did not 'remember seeing anything'.

'Did you see a telegram from the *Amerika*?'

Lightoller did not 'remember seeing any'.

'Did you see a telegram from the *Californian*?'

Lightoller did not 'remember seeing any'.

'Did you see any such memoranda?'

Lightoller did 'not remember seeing any such memorandum'.

'Was such a notation made on the chart?'

Lightoller did 'not remember seeing any myself, because I did not look'.

When asked if 'no one called your attention to any telegram or wireless from any ship warning you of ice?' he replied 'Yes', and fleshed out the story he had told before. 'I do not know what the telegram was. The commander came out when I was relieved for lunch, I think it was. It may have been earlier; I do not remember what time it was. I remember the commander coming out to me some time that day and showing me a telegram, and this had reference to the position of ice.'

'A warning to you,' Smith asked, 'of its proximity?'

'No warning,' said Lightoller, 'but giving the position – a mere bald statement of fact.' A mere bald statement of fact: it was the best description of an iceberg the inquiry had yet heard.*

Lightoller worked out that the ship would be at the position stated at around 11 p.m. and informed First Officer Murdoch. Since, asked Senator Fletcher, they knew they would be passing an iceberg that night, would it not be a sensible precaution to slow down? 'It depends altogether on conditions,' shrugged Lightoller, 'and it finally rests with the commander's judgment.'

Senator Smith digested this. Senator Fletcher then asked a vital question: following the collision, 'What was done then with reference to the ship; was her speed lessened then?'

Lightoller claimed not to know whether the ship stopped after the collision or continued on its course. 'I was below; I do not know anything about that.'

'You could not tell that?' asked Fletcher, surprised that an officer could not tell the difference between a still and a moving ship.

'I could not tell you officially,' said Lightoller. 'I know I came out on deck and noticed that her speed was lessened; yes.'

But, pressed Fletcher, 'Was she not actually stopped entirely from going forward?'

'No,' Lightoller replied, 'she was not. That is why I said, in my previous testimony, that the ship was apparently going slowly, and I saw the First Officer and the Captain on the bridge, and I judged that there was nothing further to do.'

Lightoller then made a second statement defending Ismay. He had heard from a 'reliable' witness that Ismay had been 'practically

---

*   In his memoirs, Lightoller wrote: 'Wireless reports were coming in throughout the day from various ships, of ice being sighted in different positions . . . the one vital report that came through but which never reached the bridge, was received at 9.40 p.m. from the *Mesaba* . . . the position this ship gave was right ahead of us and not so many miles distant. The wireless operator was not to know how close we were to this position, and therefore the extreme urgency of the message.' (Lightoller, *Titanic*, p. 280.)

thrown' into the lifeboat by Chief Officer Wilde (who was now dead). 'Wilde was a pretty big, powerful chap, and he was a man that would not argue very long. Mr Ismay was right there . . . and Mr Wilde, who was near him, simply bundled him into the boat.' Ismay, at six foot four inches, was a pretty big powerful chap too and Senator Smith noted that Lightoller had not remembered this incident in his previous testimony. Lightoller replied that while he had unfortunately forgotten the source of the story, he believed it to be true; Ismay, on the other hand, said that it was not true, that the decision to board a lifeboat had been his alone.

There was a transparent, seemingly deliberate, feebleness to Lightoller's defences of his employer. As a patriot he was damned if he was going to stand by and watch an Englishman savaged in this kangaroo court, but Lightoller's mockery of Smith served also to undermine Ismay. Lightoller's biographer, Patrick Stenson, claims that he 'simply felt sorry' for the boss, that he was 'one of those curious creatures' who went to the 'aid of the underdog, and certainly there was no dog more under just then than the chairman of the White Star Line'.[14] But Senator Smith did not see it like this. Baffled by the dynamic between the two men, he returned to the question, which Lightoller felt had been exhausted in his last interrogation, of when he had last seen Ismay.

'As I now recollect your testimony – and I have it here – you said you were not acquainted with Mr Ismay.'

'I have known Mr Ismay for fourteen years.'

'You did not speak to him that night?'

'I did.'

'You told me you looked at one another and said nothing.'

'I might have spoken and I might have said "Good evening".'

'I mean after the collision.'

'After the collision, no.'

'One moment,' Smith paused. 'After the collision you saw Mr Ismay standing on the deck?'

'Yes.'

'Looking out at sea?'

'I don't know what he was looking at.'

'You were standing out at deck about twenty feet from him?'

'No, sir.'

'You say now that you did not say that?'

'No, sir.'

'Would that not be true?'

'I do not think so. I was walking along that side of the deck.'

'How far past Mr Ismay?'

'I walked past him within a couple of feet of him.'

'And he said nothing to you and you said nothing to him?'

'I might have said "Good evening". Beyond that I said nothing. I had work on; something else to do.'

'Did he say anything else to you?'

'Not that I know of. He may have said "Good evening". Perhaps I said that, perhaps I did not. I do not remember.'

'In a great peril like that, passing the managing director of the company that owned the ship, you passed him on the ship and you said "Good evening"?'

'I would, as I would to any passenger I knew.'

'And he passed you and said "Good evening"?'

'I could not say. I say I may have said "Good evening" and may not, and he may have said it and he may not.'

'I only want to know as well as you can recollect.'

'I cannot say for certain.'

'My recollection is that you said you did not speak to him.'

'I am not certain. If I did speak, it was purely to say "Good evening" and nothing more and nothing less.'

'How long was that after the collision?'

'I think,' said Lightoller, 'you will find that in the testimony.'

'I know I will find it there,' said Smith, 'but I want it again. Your recollection is just a little better today than it was the other day, and I would like to test it out a little.'

'On the contrary,' said Lightoller. 'My mind was fresher on it then, perhaps, than it is now.'

According to Lightoller's granddaughter, Louise Patten, the officer confided to his wife a very different version of events. What he told her was kept a 'family secret' for nearly a century. Following the collision, when he had gone to the bridge to ask if the blow was serious, Ismay had told the Captain to continue moving 'Slow Ahead'.* The ship, which had stopped following the collision, now started up again and continued at a speed of around 5 or 6 knots until 12.15 a.m., when the Captain sent down the order to once more stop the engines. In pushing her forward, Lightoller believed, Captain Smith had allowed water to pour through the damaged hull at hundreds of tons a minute and to burst through six watertight compartments, one after another. Had the *Titanic* stood still, 'the whole ship would have assumed a fairly acute and mighty uncomfortable angle, yet, even so, she would, in all probability have floated – at least for some considerable time, perhaps all day. Certainly sufficient time for everyone to be rescued.'[15]

We cannot know whether or not Ismay gave the Captain this order, but had he done so it would not have been an unreasonable sugges-tion, and nor would it have been out of character. He was confident that the *Titanic* was unsinkable and he wanted to avoid the adverse publicity of a damaged liner being needlessly towed to port. In his testimony, Lightoller described Ismay as silent and motionless on the *Titanic* and as incapable of action on the *Carpathia*. Ismay, too, presented himself as a man who said nothing, saw nothing and did nothing. But the reason for Ismay's conflicts with his father, Wilton Oldham believed, was that Bruce was 'quick thinking', that he acted independently and made decisions without due consideration. In a crisis, when given a choice between action and inaction, Ismay was the sort of man who would always opt for action, but on a sinking ship standing still is a mark of heroism. As Kipling put it: 'But to stand an' be still to the Birken'ead drill is a damn tough bullet to chew . . . So they stood an' was still to the Birken'ead drill, soldier an' sailor too.'

* The *Titanic* contained an internal telegraph signalling system by which commands from the bridge could be shown on the engine room telegraph indicator. The positions on the clock-shaped indicator went from Stop, to Slow Ahead, Ahead One Quarter, Half Ahead, Ahead Three Quarters and Full Ahead.

'An Ismay', as journalists had noted of the family tendency, 'never goes back'. For Bruce Ismay, keeping going was better than standing still;* advancing straight at the iceberg was better than trying to swerve around it; jumping into a lifeboat was better than remaining on the ship; pushing the wrong way on an oar was better than not rowing; returning to England was better than waiting around in New York; looking forward at the horizon was better than looking back at the sinking ship. Ismay, who never again rode a horse and rarely wore an overcoat after his father's humiliations, can always be found shutting the cupboard door which contains the sea before continuing down the corridor. Jack Thayer described him in the doctor's cabin on the *Carpathia* as 'looking ahead with his fixed stare', and when he was deciding whether or not to become president of the IMM, Ismay had told Sanderson: 'I intend going slow, and giving the matter the most earnest and careful consideration.' Was Ismay's flaw that he acted too quickly, or too slowly?

It is easy to hear him give the order to Captain Smith to go 'Slow Ahead'. Ismay's refusal to believe that either the *Olympic* or the *Titanic* could sink as a result of a collision is apparent in a letter he sent on 7 March to the head of the Hamburg-America Packet Company. 'The fact there is no graving dock in America which would accommodate the *Olympic* and the *Titanic* has given me much food for thought as to what would happen in the event of one of these vessels meeting with a serious accident in American waters.'[16] It is also easy to hear the Captain – who lost control of the situation almost immediately – agreeing that to continue slowly was the right thing to do. 'I cannot imagine', Captain Smith had said of the *Titanic*, 'any condition that would cause a ship to founder. Modern shipbuilding has gone beyond that.' Why panic the passengers, most of whom were asleep, by stopping the ship? Because she had been running under a full head of steam, all eight exhausts would,

---

* Lawrence Beesley agreed. He describes how the *Titanic*, having initially stopped dead still, then 'resumed her course, moving very slowly through the water . . . I think we were all glad to see this: it seemed better than standing still.' (Beesley, *The Loss of the SS Titanic*, p. 30.)

Lightoller later said, start 'kicking up a row that would have dwarfed the row of a thousand railway engines thundering through a culvert'.[17] A few months earlier, had the *Olympic* not stayed afloat when he rammed her into the HMS *Hawke*? And here was the *Titanic*, also built like a battleship, but a thousand tons heavier.

Later, according to Lightoller's granddaughter, 'while they were still on the *Carpathia*, the chairman of the White Star Line had shown my grandfather where his duty lay. Due to certain exceptions in White Star Line's limited liability insurance policy, Bruce Ismay had told him, if the company were found to be negligent it would be bankrupted and every job would be lost. Rightly or wrongly, my grandfather decided that it was his first duty to protect his employer and his fellow employees, and in his autobiography he made it clear that this was exactly what he had done.'[18]

But if Lightoller was keeping a secret it was because he needed also to protect himself. He had been at sea for twenty-five years; he was now thirty-eight – only three years younger than Captain Rostron – with a young family to feed and he wanted his own command. How would Lightoller, who was in bed at the time, have known that the Captain was going 'Slow Ahead' under Ismay's orders? Perhaps Officer Boxhall had told him, in which case Ismay would have had to coerce Boxhall as well, but there is no suggestion that Ismay and Boxhall had any private contact whatever. And how possible is it that Ismay, dosed as he was with opiates and unable to think of anything beyond the need to delay the *Cedric* and replace his shoes, would have set in motion a full-scale operation of silencing the ship's crew? The *Titanic* would have sunk in a matter of hours whether or not the Captain had gone Slow Ahead; it was wishful thinking on Lightoller's part that had she stopped completely the ship might have remained afloat long enough for her passengers to be rescued.

But still, the suggestion remains. As Lawrence Beesley put it in an article for the *New York Times* on 29 April, in which he considered the evidence for and against Ismay's control over the speed of the ship, 'I admit the possibility, and there it must be left.'

Lightoller was followed to the stand by Quartermaster Robert Hichens, who had been at the wheel when the collision took place. Senator Smith knew that the White Star Line wanted Hichens out of the country; he had been one of the five subpoenaed men to be brought back from the *Lapland* in a US Navy pilot boat. Ismay and Lightoller listened closely as Hichens gave his evidence. He had gone to the wheel at 10 p.m. At 11.40 three gongs sounded from the look-out, followed by a telephone call 'iceberg right ahead'. Murdoch rushed to the bridge and gave the order 'hard-a-starboard'. Five minutes later the Captain entered the wheelhouse and saw, from the commutator on the front of the compass, that the ship had already listed five degrees. Why, asked Senator Smith, did you 'put the ship to starboard, which I believe you said you did, just before the collision with the iceberg?'

'I do not quite understand you, sir,' replied Hichens.

'You said that when you were first apprised of the iceberg, you did what?'

'Put the helm to starboard, sir. That is the order I received from the Sixth Officer.'

'What was the effect of that?'

'The ship minding the helm as I put her to starboard.'

'But suppose you had gone bows-on against that object?' asked Smith, who now knew where the bow was.

'I don't know nothing about that. I am in the wheelhouse and, of course, I couldn't see nothing.'

'You could not see where you were going?'

'No, sir; I might as well be packed in ice.' The spectators absorbed the image and Hichens continued. 'The only thing I could see was my compass.'

'The officer gave you the necessary order?'

'Gave me the order, "Hard-a-starboard".'

'Hard-a-starboard?'

'Yes, sir.'

'You carried it out immediately?'

'Yes, sir; immediately, with the Sixth Officer behind my back, with the junior officer behind my back, to see whether I carried it out – one of the junior officers.' Neither officer had survived.

Whether they were referring to the iceberg or to the passengers, the witnesses at the inquiry often described seeing nothing. The problem with witnesses, Senator Smith realised, was that they had been there. Their presence makes them unreliable. In an essay about the bombing of Dresden in *On the Natural History of Destruction*, W. G. Sebald wrote: 'The death by fire within a few hours of an entire city, with all its buildings and trees, its inhabitants, its domestic pets, its fixtures and fittings of every kind, must inevitably have led to overload, to paralysis of the capacity to think and feel in those who succeeded in escaping. The accounts of individual eye witnesses, therefore, are of only qualified value.'

Quartermaster Hichens was released. He had not, Lightoller also confided to his wife, told the truth. But nor had Lightoller, who, according to his granddaughter, concealed from the inquiry that when he went to Murdoch's room to collect a firearm he was told that Hichens, ordered by Murdoch to steer 'hard-a-starboard', meaning that he should turn to port, had turned the wrong way. If Hichens had indeed done this, it was an understandable error; in 1912 sailing ships and steamships operated two different steering communication systems, rudder orders for steamships and tiller orders for sailing ships, which meant the opposite of one another. Sailors who started their careers on ships which were steered by tillers connected directly to the rudder had to get used to ships now steered by wheels, and the confusion was the cause of many collisions. The *Titanic* operated tiller orders on the North Atlantic and Hichens, who had not sailed the North Atlantic before, followed rudder orders. In his panic, he did the reverse of what he should have done. The truth, Ismay apparently told Lightoller, would be the end of the White Star Line and all those who knew about it were told to say nothing. The fact that Hichens never mentioned his error throws some doubt on the story. But Ismay was clearly anxious, for reasons of insurance and reputation, that the collision be seen as an Act of God rather than human error. Lightoller saw that Hichens was put into Lifeboat 6, along with Frederick Fleet, the lookout, and Major Peuchen. The shaken quartermaster accepted the offer of some whisky and steered the tiller with his back to the sinking ship. Hichens was the only man, apart from Ismay, not to look when the *Titanic* went down.

Lightoller had initially been appointed First Officer on the *Titanic,* but found himself demoted when 'the ruling lights of the White Star Line' decided at the last minute to draft in Chief Officer Wilde, who had useful experience of the *Olympic.* 'This doubtful policy threw Murdoch and me out of our stride; and apart from the disappointment of having to step back in our rank, caused quite a little confusion.'[19] Everyone changed places: Murdoch, who had been Chief Officer, was now First Officer, Lightoller was now Second Officer, and the original Second Officer was sent home. As a Christian Scientist, Lightoller believed that death was an illusion, man was indestructible, and the power of prayer would help overcome suffering and distress. In an article for the *Christian Science Journal* he described his 'miraculous' survival and his understanding that knowledge of the 'Truth' had saved him. But as the 'solitary survivor of over fifty officers and engineers', he was now isolated; his dead friends and fellows had at least 'escaped the never-to-be-forgotten ordeal carried out in Washington'.[20]

Lightoller was caught between bedding down in a cheap hotel with an unpaid crew and conspiring with the officials of the White Star Line. First Captain Smith and now Ismay had fallen to pieces, leaving the most senior surviving officer to do his duty for the company. 'A washing of dirty linen would help no one,' Lightoller said of his performance at the inquiries.[21] He was loyal to the White Star Line but he was also a wild card, and Lightoller obeyed a higher law than the US Senate Subcommittee. He took risks, he told lies, and he changed his story when it suited him. He knew how to make words work; some of Lightoller's phrases, such as the remark that he did not leave the ship, the ship left him, would become famous (Ernest Jones reminded Freud of Lightoller's expression when he was trying to persuade the psychoanalyst to leave Vienna in 1938). An exchange with Senator Bourne about the absence of searchlights on the *Titanic* shows Lightoller – whose very name is Dickensian – to be a master of language: 'A searchlight', he explained, 'is a peculiar thing, and so is an iceberg. An iceberg reflects the light that is thrown on it, and if you throw the light on an iceberg it turns it to white, and if you throw it on the sea it turns it to white.' Lightoller could conjure up images, hold an audience in his hand. Only in *Titanic and Other Ships*, published two years before Ismay died, did Lightoller tell the full-blooded sea yarn he denied the crowd in America. No longer the silent movie he described at the inquiry, Lightoller now turns up the sound and lets us hear the night. It was impossible, he reveals, to give orders to the crew and the passengers because of 'the appalling din', the 'infernal roar', of the steam being released when the engines stopped. Lightoller had to cup his hands to the Captain's ear and shout to ask him if the boats should be put out. After half an hour, when the ship is still and the noise has ceased, 'there was a death-like silence a thousand times more exaggerated'.[22]

On 25 April, Ismay replied, in pencil, to a letter he had received from the brother of the artist, Francis Millet, who had gone down with the ship. 'I regret extremely', he said, that 'I had not the pleasure of meeting your brother and am therefore unable to give you any information in regard to him.' Ismay expressed his condolences, adding that 'I would also like to say that I am not in any way responsible for the truly dreadful disaster which God knows, if it had been possible, I would have done anything in the world to avert. We had the finest ship in the world, and appointed as her commander a man in whom we had absolute confidence. Why people try to make me responsible for the horrible disaster I cannot imagine. The last week has been a horrible nightmare to me, and I cannot yet realise the *Titanic* has gone. I can only hope God will give me strength to see the matter out to the end. I am having a truly awful time.'[23]

Later that day Ismay wrote to Senator Smith reminding him that 'though under severe mental and physical strain' he had 'welcomed

this inquiry' and placed himself at the disposal of the committee.
'Though not in the best of condition to give evidence', he had 'testified at length' the previous Friday. Since then he had attended every
hearing and held himself in readiness to answer all the committee's
needs, 'though personally I do not see that I can be of any further
assistance'. Might it be possible, he implored, 'if the committee
wishes to examine me further', for it to do so 'promptly in order that
I may go home to my family', especially now that the British government had begun its own inquiries 'which urgently require my
personal attention in England'?

Smith responded immediately. 'I can see that your absence from
England at a time so momentous in the affairs of your company
would be most embarrassing, but the horror of the *Titanic* catastrophe and its importance to the people of the world call for scrupulous
investigation into the causes leading up to the disaster . . . I am
working night and day to achieve this result, and you should continue
to help me instead of annoying me and delaying my work by your
personal importunities.' In London, an editorial in the *Saturday
Review* asked: 'Why should British subjects be detained against their
will . . . in order that a blustering ignoramus may tease them with
questions about the difference between the "bow" and the "head" of
a ship, the origin of icebergs and the use of watertight compartments?' Senator Smith afterwards declared that 'energy was more
desirable than learning', and 'If I asked questions that seemed absurd
to sailors, it did no harm. Everybody isn't a sailor, and lots of people
who have never been to sea want to know all about the loss of the
Titanic, even down to the inconsequential details that the marine
experts scoff at. And I *know* we got the truth.'

During the week, Ismay had listened as the inquiry span itself
away. He seemed like a spider trapped in an enormous web; his
simple story now had sidetracks which led nowhere, cross tracks
which started back and began anew. He was entangled, but he could
also see more clearly. He gained a fuller sense than ever before of the
world inside his ships; he got to know his employees, his passengers
and the names of the crew who had died; he heard how the firemen,

engineers and Marconi operators had never left their posts; he learned how wireless worked, how on the port side Lightoller was operating a 'women and children only' policy while on starboard they were loading women and children first, and he was forced to imagine what the ship had looked like when she went down. He heard that the *Californian,* a fellow ship in the IMM combine, had been eight miles away all night with her wireless turned off. He heard accounts of miraculous survivals, and he re-imagined his own death again and again. Did the hundreds who died see, as drowning figures are meant to do, their lives pass before them? Would that not have been a better fate than a survival with no future? Had he gone down with the ship, what would the papers be saying about him now? Who then would be their scapegoat? Ismay understood that every person saved was seen to be taking the place of someone else who had more right to a life. He learned that no one, not even women and children, was allowed to survive the *Titanic.*

What constituted a child anyway? Lightoller had ordered Emily Ryerson's son, a boy of thirteen, to stay on board with the men; the child only joined his mother in the lifeboat when his father, a power-ful steel magnate, intervened. 'No more boys', a frustrated Lightoller was heard to mutter. Some men felt that the women who owed their lives to male gallantry were hypocrites, and witness after witness described how wives and mothers in the lifeboats had refused to return to their dying husbands and sons. Real wives were those like Ida Straus, who remained on board with her husband of fifty years, the owner of Macy's department store. 'Where you go, I go,' Mrs Straus reportedly said. But Mrs Stuart White, a first-class passenger who lived at the Waldorf-Astoria, believed the women in the life-boats had been braver than the 'heroes' who only stayed on the ship because they thought it was the safer option. 'I do not think that there was any particular bravery,' Mrs White told the inquiry, 'because none of the men thought it was going down.'

By the end of the inquiry, Ismay knew more about those final hours than anyone else who had been unlucky enough to survive the *Titanic.*

The crewmen were finally released on 29 April, but Ismay was not. The next day the *Mackay Bennett,* contracted by the White Star Line to search for bodies, returned to Halifax. They had found 306 of the 1,500 dead. In the opinion of the surgeon who accompanied the 'funeral ship', most had lived for four hours in the water before freezing to death. Those who could not be identified were buried at sea, and 190 others, including the body of John Jacob Astor, were brought home.

When Ismay was finally called to the stand for the second time on 30 April, Senator Smith found himself addressing an apparently contrite and more responsive man. Perhaps Ismay was relieved: no one else from Collapsible C had been cross-examined, not even William Carter, who had left the ship with him, or Quartermaster Rowe, who had taken charge of the boat and knew whether or not Ismay had been rowing; nor had Dr McGhee been called to answer questions about Ismay's state of mind while on the *Carpathia.* The inquiry was drawing to a close and apart from the claims that Emily Ryerson had made in the press, which she was due to swear in an affidavit the following day, Senator Smith had not been able to pin a thing on Ismay. He understood that speed was of less importance to the White Star Line than luxury, that Ismay did what he could to lower and load the boats on the starboard side, and that whether he jumped into Collapsible C or was pushed, he was in one of the last boats to leave. It was tacitly agreed by the Senate Committee that Ismay had done his duty, without anyone's knowing quite what that duty was.

Most of what Smith now asked him mirrored what he had asked before. Having questioned Ismay about the structure of the IMM, the relationship of the White Star Line with the Morgan combine and Harland & Wolff, and their arrangement with the British government to deliver the mail, Smith asked Ismay to explain once again his role on the *Titanic.* Was he there officially for the purpose of inspecting? Ismay replied that he had not yet *made any inspection of the ship at all*, that during the voyage, he *was never outside the first-class passenger accommodation on the ship*. Ismay repeated that he had not dined with the Captain on Sunday night, that the Captain had been at a dinner for the ship's most prominent figures organised by the Wideners (to which Ismay, despite knowing George Widener, was not invited). He explained that the ship was insured for $5 million and expressed horror at the suggestion that an attempt might have been made by him to reinsure the vessel on Monday 15 April. It was *a horrible accusation* and Ismay would have considered any such action entirely *dishonourable*. He was asked to read aloud the same Yamsi messages read by Franklin the week before, and to give the time and context of each. He was asked to confirm the speed of the ship, and to say whether 'anyone had urged the Captain to greater speed' than seventy revolutions. *It is really impossible*, said Ismay with disgust, *to imagine such a thing on board ship*. He replied, when asked if he did not 'regard it as an exercise of proper precaution and care to lessen the speed of a ship crossing the Atlantic when she had been warned of the presence of ice ahead' that it was *a question I cannot give any opinion on. We employ the very best men we possibly can to take command of these ships, and it is a matter entirely in their discretion.*

Ismay told the inquiry that following the collision Captain Smith gave no report as to the extent of the damage and, as far as he was aware, raised no alarm. The information Ismay received – that the ship had thirty-five or forty minutes to live – came from Joseph Bell, the Chief Engineer, after he had seen the Captain on the bridge. However, Ismay stressed that *Captain Smith was a man with a very very clear record. I should think very few commanders crossing the Atlantic have as good a record as Captain Smith had, until he had the unfortunate incident with the Hawke.* Had the *Olympic's* earlier collision with HMS *Hawke* shaken Ismay's confidence in the Captain? He replied that it had not and that he had *no reason to doubt* that Captain Smith had *quite got over* the mishap. As to when the Captain gave Ismay the Marconigram from the *Baltic*, he stated *I do not know whether it was in the afternoon or immediately before lunch; I am not certain. I did not pay any particular attention.* How did it happen that the *Titanic* had only twenty lifeboats? *That was a matter for the builders, sir, and I presume that they were fulfilling all the requirements of the Board of Trade*. Ismay agreed that the Board of Trade requirements were out of date, but when asked to confirm whether 'the davits they had on the *Titanic* were capable of handling three boats instead of one and that there was no question about those davits being able to handle twice the number of boats they did handle', Ismay refused to answer. *I could not express an opinion in regard to that. I do not know anything about it*. As to whether the lesson of the *Titanic* showed that it was 'impossible to construct a non-sinkable ship', he said that he *would not like to say that, because I have not sufficient knowledge*. This time when Ismay was asked whether his lifeboat was filled to its capacity, he replied that it was not, that *the full capacity of one of those boats is about sixty to sixty-five*. Who told him to enter that lifeboat? *No one, sir*. Why did he enter it? *Because there was room in the boat. She was being lowered away. I felt the ship was going down, and I got into the boat*.

Ismay was released and there were expressions of courtesy all around. Should he be needed by Senator Smith in the near future, he *would be quite glad to come back from the other side*. In the audience that day had been John Galsworthy. 'A queer, jumbled business,' Galsworthy wrote in his diary. 'We heard the unfortunate Ismay give his evidence very quietly and well. The system and public is to blame for the miserable calamity; and of course the same public is all agog to fix the blame on some unhappy shoulders.' Now that the inquiry was drawing to a close, the life of the city was resuming. 'To the *Trovatore* at the Opera,' Galsworthy continued. 'Phew! What a rococo!' Unable to return on the *Titanic* as planned, Galsworthy sailed to England five days later on the White Star liner the *Baltic*, whose ice warning Ismay had carried around in his pocket all day. 'Reading; Talking; Eating; Sleeping,' he wrote of the voyage back. 'Nothing eventful happened.' [24]

Ismay caught the train to New York, ignoring summonses issued by the District Supreme Court to testify in an action brought by Mrs Louise Robins, the widow of John Jacob Astor's valet. When Ismay did not appear, the hearing was abandoned.

Senator Smith too returned to New York where, at the Waldorf-Astoria, he took affidavits from various survivors including August Weikman, the *Titanic's* barber, who was probably the source of Lightoller's story that Ismay had been 'ordered' by the officer in charge into 'the last boat to leave'. Ismay's actions were justified, Weikman swore, because 'there were no women in the vicinity'. Mrs Malaha Douglas, travelling first-class, claimed that on the afternoon of the accident she and her now-deceased husband had seen a seaman take the temperature of the water not by lowering a pail down the side of the ship, but by filling the pail with water from a tap and then placing his thermometer inside. Mrs Douglas also claimed that Ismay had been at the Wideners' dinner party that night, and repeated the story told by Emily Ryerson of his showing an ice warning to Mrs Ryerson and Mrs Thayer that afternoon, before announcing that they would deal with the problem by lighting more boilers.

Emily Ryerson then arrived to give her affidavit. Smith's key witness described her last hours on the ship but said nothing about the now famous encounter with Ismay. Why she held back would be clearer to Ismay than it was to Smith, but she doubtless said that it was to avoid additional stress. Mrs Ryerson, who had been returning on the *Titanic* for the funeral of her eldest son, was now also burying her husband. If Lightoller, who tried to prevent her thirteen-year-old boy from boarding the lifeboat, had had his way, both her sons would now be dead. Getting tangled up in what Ismay did or didn't say about speed was more than she could bear this week. Marian Thayer, who had been with Mrs Ryerson during the encounter, had said nothing whatever about it. She was caught between loyalty to an old friend and a new one.

On 2 May, Ismay left for England on board the *Adriatic*. He was, he told a British journalist, 'worn out. I am feeling very tired and wish to retire.' The previous day, the funeral of John Jacob Astor had been knocked off the front pages by the appearance of 15,000 suffragettes parading through New York.

The Senate's report of its findings was published on 28 May. It seemed that for all his support of the 'little man', Smith was uninterested in the part played by class on the *Titanic*. He saw the conflict as between good men and bad, rather than rich men and poor. Smith duly noted that 60 per cent of first-class, 42 per cent of second-class and 25 per cent of third-class passengers were saved, but he questioned only three steerage passengers, each of whom assured him that those in third-class had not been trapped in the ship's bowels.

William Alden Smith's aim, 'plain and simple', had been 'to gather facts relating to this disaster while they were still vivid realities'. The facts he gathered were these: that the *Titanic* was fitted with davits that could hold forty-eight lifeboats but which were carrying only sixteen (the number approved by the British Board of Trade); that the lads in the crow's nest were not provided with binoculars; that while the Captain was at a private dinner party made up of East Coast millionaires the ship was steaming at 22 knots – her maximum speed of the voyage – into an ice field which he had been warned about by Marconigram on many separate occasions; that after the collision the *Titanic* continued to go forward for thirty-five minutes; that the Captain failed to inform his passengers and crew that the ship was sinking; that he gave no instructions as to the handling and manning of the lifeboats, and there was no order to abandon ship. The single fact that Senator Smith had not been able to ascertain, despite repeatedly asking him, was one that his star witness himself was hardly able to understand: why had Ismay jumped?

The language William Alden Smith used in his report to describe these facts was neither 'plain' nor 'simple'; he was writing a florid page of history. Six weeks after its loss, with the mariners still wet, the *Titanic* had become an historical document. Senator Smith had gathered together myriad witness accounts, all of them partial, some false, and allowed them to interface, interlace and interfere. He had thus given the chaos of the night a start, a middle and an end. There was a beauty to the process of reconstruction, and the romance of his role was not lost on him. The Senator concluded that 'the willing-ness' of Captain Smith 'to die was the expiating evidence of his fitness to live . . . In his horrible dismay, when his brain was afire with honest retribution, we can still see, in his manly bearing and his tender solicitude for the safety of women and little children, some traces of his lofty spirit when dark clouds lowered all about him and angry elements stripped him of his command.' He spoke of the 'bosom' of the ocean, the 'glamour of the sea', and 'the daring spirit of the explorer and the trader' whose noble 'calling' was 'already demoralised and decadent'.

And so it rolled on, page after page of high oration, until he came to the part played by Ismay. Here William Alden Smith showed himself to be a man of hidden dimensions. Ismay's personal conduct, on which the inquiry had expended so much energy, was not discussed, and nor did Senator Smith comment on the chairman's survival. Instead, he shifted the debate onto an entirely different level. While it could not be proven that Ismay had ordered the Captain to keep up the ship's speed, Senator Smith concluded that the 'presence of the owner *unconsciously* stimulates endeavour'. Did he mean that Captain Smith was unconsciously stimulated by Ismay or that Ismay had unconsciously stimulated the Captain? Whose unconscious was Senator Smith referring to? Was the Captain acting as Ismay's unconscious, or Ismay acting as the Captain's?

Smith had chosen a word the impact of which was only quietly beginning to make itself felt in America. In Vienna, Freud was conducting his own inquiries into the unconscious which, he discovered, ruled everything. What distinguished the hysteric, Freud believed, was the 'inability to give an ordered history of their life', a narrative he compared to 'an unnavigable river whose bed is now obstructed by masses of rock, now broken and made shallow by sandbanks'. Moments of clarity and coherence are followed by the drying-up of information, 'leaving gaps and mysteries'. Connections are fragmented, the sequence of events becomes uncertain, the speaker will correct a fact or a date and then, 'after a lengthy vacillation', return to the original statement. The reasons behind such a disordered narrative, Freud suggested, could either be 'deliberate dishonesty' – 'the patient is consciously and deliberately holding

back a part of something that is very well known to her' – or 'unconscious dishonesty', the innocent result of amnesia or repression.[25]

The concept of the unconscious was not Freud's alone. It was he
who distinguished it as a place separate from consciousness as
opposed to a state into which you fall, but Freud always said that the
unconscious had been 'discovered' by 'the poets and philosophers
before me'.

Joseph Conrad, like Senator Smith, had no knowledge of psycho-
analysis but knew there was a link between the psyche and the sea.
When Senator Smith spoke of Ismay's pressure on the Captain as
'unconscious' he was drawing, as Conrad often did, on the early
twentieth-century interest in *doubleness*: the conscious self we think
we know is controlled by a powerful stranger who inhabits us
unawares, who entangles himself in our speech, who pushes us
forward and holds us back. Duality runs like a fever through an
astonishing number of novels and stories in the late nineteenth and
early twentieth centuries, particularly those of Henry James, who left
his native Albany to live in England: in 'The Jolly Corner' (1908),
Spencer Brydon leaves New York for Europe and returns to the
family home thirty years later to discover it haunted by the self he
might have been had he stayed in the city of his birth. Oscar Wilde's
The Picture of Dorian Gray (1891) dealt with the 'terrible pleasure of
the double life', and in Robert Louis Stevenson's *Strange Case of Dr
Jekyll and Mr Hyde* (1886), Hyde 'drew steadily nearer the truth, by
whose partial discovery I have been doomed to such a dreadful shipwreck: that man is not truly one, but truly two'. In 1913, the poet
T. E. Hulme described, in 'Speculations', the self as being composed
of two parts, a 'superficial self' and a 'fundamental self': it is the
fundamental self, he said, that 'leaps into action'.

PART II

On Land

And as the smart ship grew
In stature, grace, and hue,
In shadowy silent distance grew the Iceberg too.

Alien they seemed to be;
No mortal eye could see
The intimate welding of their later history,

Or sign that they were bent
By paths coincident
On being anon twin halves of one august event,

Till the Spinner of the Years
Said 'Now!' And each one hears,
And consummation comes, and jars two hemispheres.

Thomas Hardy, 'The Convergence of the
Twain (Lines on the Loss of the *Titanic*)'

The Convergence of the Twain

I had jumped . . . it seems . . .

Joseph Conrad, *Lord Jim*

If we each have an author who is perfectly equipped to tell our tale, Joseph Conrad would be Ismay's, and for a brief moment he was. Surrounded by newspapers in his house near Ashford in Kent, Conrad watched 'the luckless Yamsi', as he called him, begin his long descent. 'This affair of the *Titanic* has upset me,' he told his agent, 'on general grounds, but also personally. I am not doing well.' Conrad loathed the 'festive' air of the press as they celebrated the 'heroism' of the dead, and the satisfaction of the Americans that 'this fatal mishap should strike the prestige of the greatest Merchant Service of the world'. But, with the loss of the manuscript of his story 'Karain: A Memory', which he was selling to John Quinn, the American collector of modernist writing, part of Conrad's own life too had gone down with the ship. Because he had not insured the package, Conrad was now £40 out of pocket. 'I depended on that sum,' he complained.[1]

Twelve years earlier in October 1899 – the month before Ismay took over the chairmanship of White Star Line – Conrad had written a despairing letter to Ted Sanderson, the son of the Reverend Lancelot Sanderson, Ismay's former headmaster at Elstree.

My dear Ted, You have much to forgive me: but try to imagine yourself trying your hardest to save the School (God forefend) from downfall, annihilation, and disaster: and the thing going on and on endlessly. That's exactly how I am situated: and the worst is that the menace (in my case) does not seem to come from outside but from within: that the menace and danger or weakness are in me – in myself alone . . . I fear! I fear! . . . I am now trying to finish a story which began in the Oct. No. of *Blackwood.* I am at it day after day, and I want all day, every minute of a day, to produce a beggarly tale of words or perhaps to produce nothing at all. And when that is finished . . . I must go on, even go on at once and drag out of myself another 20,000 words, if the boy is to have his milk and I my beer (this is a figure of speech – I don't drink beer, I drink weak tea, and yearn after dry champagne) and if the world is not absolutely to come to an end.[2]

Being menaced by an internal danger or weakness – they are the same thing for Conrad – is his recurring theme, and again and again his writing reduces him to this condition of anguish. What he is forcing out of himself is a tale of a man who jumps from a sinking ship and lives on with 'the acute consciousness of lost honour'.

Jim, the son of a country parson, has a sense of maritime heroism born from 'a course of light holiday literature'. He immerses himself in yarns of pirates and poop decks, crow's nests and compasses, sailing ships and savages. He dreams of the ancient chivalry of the sea, he yearns for the endlessness of the horizon. He joins the Mercantile Marine where he proves himself 'gentlemanly, steady, tractable', and then takes a berth as chief mate on the *Patna,* a rusty Chinese-owned, Arab-chartered steamer, 'worse than a condemned water tank', carrying 800 pilgrims across the Indian Ocean from Singapore to Mecca.

Jim's world has become a ship, and he is the hero of his own adventure. Standing on the bridge of the *Patna* he watches the night descend 'like a benediction'; he marvels at the 'assurance of everlasting security', the unbounded safety and peace shed from the rays of the stars. In the excess of his wellbeing, he knows there is no noble deed he will not do, no challenge he cannot face. Then, inexplicably, there is an accident of some sort. A faint noise, less than a sound, no more than a vibration, passes slowly beneath the steamer like a rumble of distant thunder and the ship quivers in response: 'suddenly the calm sea, the sky without a cloud, appeared formidably insecure in their immobility, as if poised on the brink of yawning destruction.' At the subsequent inquiry Jim will say, 'She went over whatever it was as easy as a snake crawling over a stick.'

The *Patna* begins to lean; 800 passengers are sleeping: on mats, on prayer carpets, on rough blankets, on bare planks, on decks, in dark corners all over the ship which the crew now believe will sink. There are only seven lifeboats; it is not possible to save everyone and so the Captain, a vulgar and obese German, decides to abandon ship with three of his equally shoddy officers, leaving the human cargo to their fate. The sea is as 'still as a pond, deadly still, more still than ever sea was before'; the conditions are 'rare enough to resemble a special arrangement of malevolent providence' and the crew are struggling like lunatics to release the lifeboat without waking the pilgrims.

Standing apart from them all, Jim has not yet been tested 'by those events of the sea that show in the light of day the inner worth of a man'. The crew are animals, he has always known that – 'those men did not belong to the world of heroic adventure'. Jim has always seen himself as separate, as singular; he recognises his superiority. The time has now come to prove himself the man of predestined courage he feels himself to be. This is the moment he has been preparing for since he was a child. But instead of rising to the well-rehearsed occasion, instead of taking the situation in hand, Jim does nothing, says nothing, he has no idea what to do: instead he stands stock-still in a daze on the starboard side of the bridge while the crew struggle with the lifeboat on the port side, expecting at any moment the sea to submerge them all. Should he cut down the other lifeboats so they can float off the ship when she eventually founders, giving some of the passengers a chance to live? Should he wake the pilgrims to tell them that they are about to die? 'Where was the kindness in making crazy with fright all those people I could not save single-handed – that nothing could save?' he reasons. Jim delays acting; he is paralysed: all he can think is 'eight hundred people, seven boats and not enough time, eight hundred people, seven boats and not enough time'. 'You think me a cur for standing there, but what would you have done?' he later asks.

The renegade lifeboat is dropping down to the water and the absconding crew, in their sleep suits, are shouting up not for Jim, but for their friend George to join them. George, on the deck, falls down dead from a heart attack. The men in the boat below are calling 'Jump, George! Jump! Oh, jump!' It is pitch black, there is a squall approaching; the *Patna* starts to plunge, and suddenly Jim moves. 'Something had started him off at last, but of the exact moment, of the cause that tore him out of his immobility, he knew no more than the uprooted tree knows of the wind that laid it low.' What happens next is against his conscious volition. 'I had jumped . . . it seems,' he recalls. He jumps from a height he can never scale again, he jumps into 'an everlasting deep hole', and as he jumps he begins to unravel.

Whatever pushes him off the *Patna*, Jim now knows that he is not and never will be the man who 'saw himself saving people from sinking ships, cutting away masts in a hurricane, swimming through surf with a line . . . always an example of devotion to duty, and as unflinching as a hero in a book'. As the lifeboat pulls away, he listens for the cries of 800 people being 'pounced upon in the night by a sudden and violent death', but hears nothing. There is nothing but silence coming from the wreck, and it is too dark to see it go down. It is too dark for the crew to see that it is not George sitting with them in the lifeboat but Jim, and when they realise their friend has been replaced – they do not yet know that George is dead – they accuse Jim of being a coward, and 'too much of a bloomin' gentleman' to help to lower the lifeboat which has now saved his life. 'Come out of your trance did you?' one of the engineers mocks, 'to sneak in? I wonder you had pluck enough to jump. You ain't wanted here.' Nor does Jim want to be there; he is not one of them. He thinks about jumping again, this time off the lifeboat and swimming back to the site of the ship to drown alongside the pilgrims, but

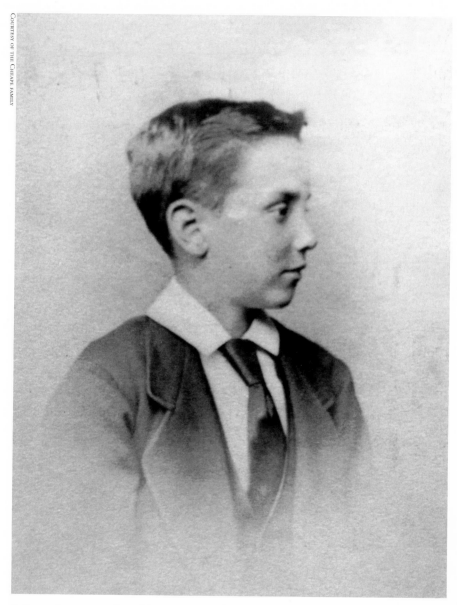

J. Bruce Ismay was the odd one out in a family of doubles. Sandwiched between two dead siblings, he was succeeded by two sets of twins.

The first of his family to receive the education of a gentleman, in 1874 Ismay went aged eleven to Elstree in Middlesex, to prepare him for Harrow. A sea-gazing northerner in the south of England, he was miserable at school.

In 1882, J. Bruce's father Thomas Ismay commissioned Richard Norman Shaw to design him a palace on 390 dank acres overlooking the River Dee. Florid at sea, Ismay senior was austere on land, the absence of trees and flowers emphasising the severity of the family home.

The Ismay family, including spouses and twins. *Seated centre, then clockwise*: Margaret Ismay, Charlotte, Bower, Ethel, James, Lady Margaret Seymour (James's wife), Bruce, Thomas Ismay, Florence, (unidentified gentleman), Dora and Ada.

The picture room at Dawpool, whose domed ceiling was described by Ismay's mother, Margaret, as 'an eyesore', served as a gallery for Thomas Ismay's art collection, which included Rossetti's *The Loving Cup* (on the easel to the left).

In New York, where he worked as an agent for the White Star Line, Ismay behaved much as any rich, handsome, unattached 22-year-old male would in a city four thousand miles away from his oppressive father.

Following their wedding on New York's Fifth Avenue, Ismay and his new wife Florence sailed to Liverpool on the *Oceanic* to meet his family.

Margaret, the Ismays' eldest daughter, was born in 1889. Following the death of their second child and the cooling of their marriage, Margaret became Ismay's favourite.

Sandheys, outside Liverpool: the Ismays' family home. Ismay, who liked order and quiet, had a separate wing built in which to house the children.

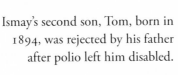

Ismay's second son, Tom, born in 1894, was rejected by his father after polio left him disabled.

Workers leaving Harland and Wolff in Belfast, 1911. The shipyard covered 80 acres, employed 16,000 local men and distributed £28,000 in weekly wages. The *Titanic* can just be seen under construction in the gantry behind.

Draftsmen designing
the White Star Liners
in the drawing room
at Harland and Wolff,
1908.

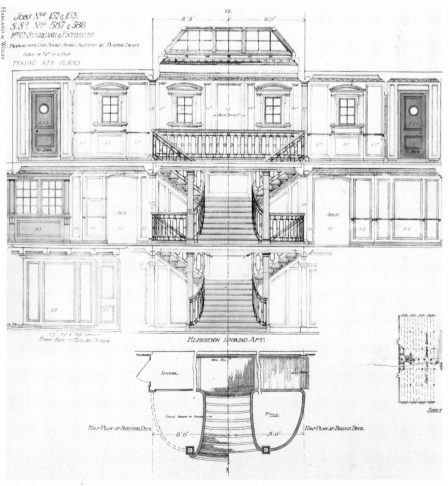

The designer's internal drawings for the *Titanic*'s first-class staircase.

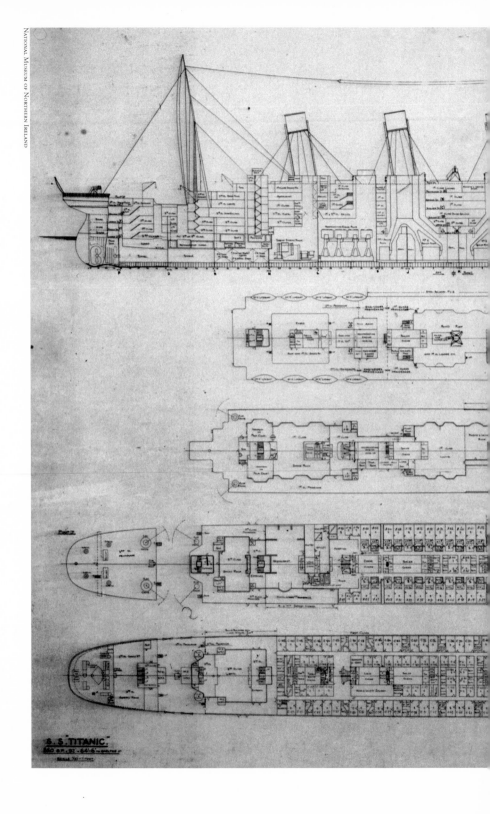

S.S. TITANIC

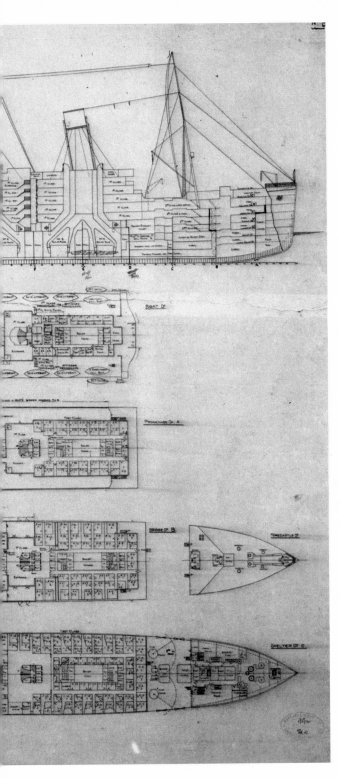

Plans for the construction of the *Titanic*. At 882 feet 9 inches in length, 92 feet 6 inches in width, 175 feet high and weighing in at 46,328 tons, the *Titanic* was the largest ship ever built, and at her launch, the largest moving object on Earth.

The *Titanic* carried four funnels – although the fouth was a dummy, thought by the designers to make the ship look more powerful. Each was the height of a two-storey house and wide enough for two locomotives to pass through simultaneously.

The *Titanic* and *Olympic* in the gantries. 'For months and months,' wrote Filson Young, 'in that monstrous iron enclosure there was nothing that had the faintest likeness to a ship; only something that might have been the iron scaffolding for the naves of half-a-dozen cathedrals laid end to end.'

Prior to her launch, Lord Pirrie and Ismay (*right*) inspect the completed *Titanic* at Harland and Wolff.

'The trouble with the beds,' Ismay noted in his inspection of the *Olympic*, 'is entirely due to their being too comfortable.' With coal in the grates and curtains on the windows, those in first class were able to feel that they were in their own private apartment.

The two first-class entrance staircases on the *Titanic* were the ship's most luxurious fittings. Built in the English style of William and Mary, the iron banister grillwork was inspired by the French court of Louis XIV. Clad in oak panelling with bronze cherubs supporting the ornamental lamps, a large glass dome allowed in natural light.

'The reading and writing room is of 1770, but in pure white, with an immense bow window', as one journalist described it.

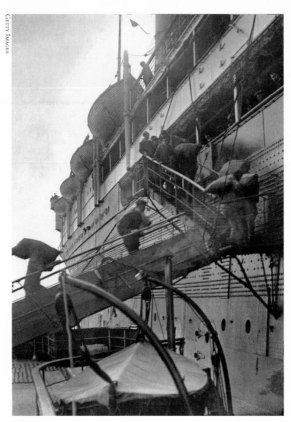

Loading the mailbags prior to departure from Queenstown (Cork), Ireland. Among the 200,000 letters and packages carried by the *Titanic* was the manuscript of Joseph Conrad's story, 'Karain'.

The *Titanic* leaving Queenstown on 11 April 1912, her last port of call before crossing the Atlantic.

The *Titanic*'s boat deck. It was Ismay's decision to restrict the number of lifeboats on the davits so that the deck was not cluttered.

Marconi room on the *Olympic*, as the *Titanic*'s would have looked. Guglielmo Marconi's recent invention of wireless messaging allowed ships for the first time to communicate with one another without using flags and flames. Marconigrams, it was believed, made the sea as safe as a suburban street.

The *Titanic*'s captain and officers. Captain Smith is seated, second from right.

Contemporary artist's impression of the *Titanic* at night. 'To stand on the deck of the *Titanic*,' Lawrence Beesley later said, 'gave one a sense of wonderful security.'

calms himself with the thought that it will be too late. Around him the crew discuss what they have just done 'as though they had left behind them nothing but an empty ship'.

A 'mysterious cable message' then arrives in Bombay. It contains an 'ugly fact', a ghastly joke: by some extraordinary chance, the *Patna* did not go down but was rescued and towed by a French gunboat to Aden with all her passengers alive, by which point the crew, now on shore, have reported that she 'sank like lead'. The result is a maritime scandal, and the story will become legendary, a topic of debate in every port and harbour for years to come.

The official inquiry into the case of the *Patna* is held in August 1883 in a police court in Bombay. Because the Captain and the crew have once again fled, Jim is the only one left to appear in the witness box. 'I might jump,' he says, 'but I don't run away', and he stands there defiant; he has done nothing, he tells himself, of which to be ashamed. Crowds fill the courtroom, spellbound by the tall, young, white man; everyone connected with the sea is here, no one has talked of anything but the *Patna* since the incident became known. They have turned up today not to discover how the ship was damaged; no one is interested in the ship herself – it is assumed that she went over some submerged wreck. They are here to see someone 'trying to save from the fire his idea of what his moral identity should be'.

The inquiry is the first time Jim has spoken since his jump, and when he answers the questions put to him, questions aiming at facts – 'as if facts could explain anything!' – the words he utters appear meaningless to him. Jim feels that he will never speak again. He is tempted to cry out: 'What's the good of this, what's the good!' when, amongst the myriad faces on the benches below, he catches the intelligent, interested gaze of Captain Marlow, who alone seems aware of the young man's struggle. Marlow asks him to dinner at the Malabar Hotel that night where, over coffee and cigars, Jim exclaims: 'I would like somebody to understand – somebody – one person at least! You! Why not you?' He can never go home now, Jim says; the scandal will have been in all the papers and his father, who has high ideals and fixed moral standards, will not understand. A few days before he

boarded the *Patna*, Jim had received a letter from his father instructing him, from the 'inviolable shelter of his book-lined, faded and
comfortable study', not to give way to temptation, never 'do anything
which you believe to be wrong'. The parson's quiet corner of England
is as clear and innocent as a child's gaze. Jim has for ever exiled
himself but he cares nothing for that. He is obsessed, Marlow realises, not by what he has lost but by the immensity of what he would
have gained had he stayed on board the *Patna* and become the hero
of the hour: 'Ah! what a chance missed!' Jim cries, 'My God! What a
chance missed!'

While Jim wants only to get away from his wretched story,
Marlow's interest in his case has just begun. Jim is the loneliest man
in the world but he is also, Marlow sees, 'symbolic', and he closes in
on Jim's consciousness like a surgeon with a scalpel. He subjects it to
the last analysis, he turns it around and around and inside out, he
holds it upside down, he takes it apart, he approaches it head on,
askance, up close, from a distance; he looks at Jim's lost opportunity
from every possible angle, examines Jim's future prospects in every
available light, wrings out each emotion, gathers alternative perspectives. He weighs to the last scruple Jim's noble intentions and
balances them against his feeble performance, he weeds out what
has lain below, unwatched and half-suspected 'like a snake beneath
a stone', and envelops the whole in a language of exquisite subtlety
and precision. Marlow alone sees that when Jim jumps from the
Patna he confronts, for the first time, himself. 'I had jumped . . .
it seems', Jim says; his jump is a non-jump: a movement of a muscle
took place but Jim was not aware of it. It was George and not Jim
who was supposed to jump but George had dropped down dead,
and for an instant Jim identified himself with the dead man and did
what *he* would have done. Something inside him had jumped, while
Jim himself remained still. For the rest of his life, and for the rest of
the book, Jim is exorcising, while Marlow examines, this stranger
within.

Lord Jim, as Conrad finally called his tale, began life in April 1898 as 'Jim: A Sketch'. It was based on an incident which became the focus of interest in Singapore in 1880: the SS *Jeddah*, carrying 950 pilgrims to Mecca, sprang a leak and was abandoned by her crew. The officers reported the ship lost, to then hear that she had been towed to Aden with all her passengers alive. The scandal became the subject of an inquiry in Aden and a debate in the Singapore Legislative Assembly. The *Straits Times* reported in September 1880 that 'public excitement has risen to fever pitch' in 'surveying the conduct' of the *Jeddah*'s captain and crew. The story was also vividly and extensively covered in the English newspapers read by Conrad in London as he was waiting for a passage to Sydney. 'DREADFUL DISASTER AT SEA: LOSS OF NEARLY 1,000 LIVES', ran the *Globe*'s headline when it was still believed that the *Jeddah* had sunk. 'We trust that no Englishman was among the boatload of cowards who left the *Jeddah* and her thousand passengers to fend for herself,' wrote the *Daily Chronicle* when the truth was revealed. The crew, Conrad believed, had betrayed 'a tradition . . . as imperative as any guide on earth could be'. But it was the possibility of betrayal, the proximity we all have to failure by which he was fascinated.

'I always suspected', Conrad said, that 'I might be no good.' He took as his model for Jim the *Jeddah*'s chief mate, Augustine Podmore Williams, the strapping young son of an English country parson who had, like Jim, started as a cadet on a training ship. Williams, who claimed that he did not jump but was 'thrown overboard' by the pilgrims, was severely condemned by the inquiry but stayed in the East and 'worked out his salvation' as a water clerk, marrying a sixteen-year-old Singaporean and fathering sixteen children. He faced out his crime as a gentleman should, and it was the manner of his living on which interested Conrad as much as the loss of his honour. In his preface to *Lord Jim*, Conrad writes that 'one sunny morning, in the commonplace surroundings of an Eastern roadstead', he saw Augustine Podmore Williams 'pass by – appealing – significant – under a cloud – perfectly silent. Which is as it should be. It was for me, with all the sympathy of which I was capable, to seek fit words for his meaning. He was "one of us".' The biographer Norman Sherry is convinced that Conrad not only saw but spoke to Williams, and 'heard his history from the man himself. I feel certain that it was his intimate knowledge of Williams's life and character, in fact, which led Conrad "to seek fit words for his meaning" with all the sympathy of which he was capable.'[3]

Conrad had been working as a writer for four years when he began his sketch of Jim, which was to be a short story of 20,000 words and completed, he anticipated, by April 1899. But he wrote *Heart of Darkness* that year instead, and when the April deadline passed for the Jim story the submission date shifted to July, and then August. 'I am utterly weary of thinking, of writing, of seeing, of feeling, of living,' Conrad complained to John Galsworthy in September. The first four chapters of *Lord Jim* were published that October in the literary monthly *Blackwood's* and Conrad thought that maybe another four instalments would be enough to round the thing off. But in November he revised his opinion: the book would be complete in five instalments; it would be twelve chapters long and ready by the end of December. The New Year dawned and Conrad had now written eighteen chapters; he would finish by the end of the month. By February he had completed twenty chapters and was no longer forcing the words out of himself. 'It comes! *it* comes', Conrad cried; the writing was taking him over, the book was writing him. He had thrown himself down a building with no ground floor. A 20,000-word story had doubled in length, then doubled again, then again. It was changing shape daily; Conrad was describing, as the reviewer for the *New York Times Book Review* put it, everything in three dimensions. He was greedy for words, he piled them high and stretched them out, he loaded the sentences down, stuffing them to the limit like bags which had to be got across the room before they burst apart. In April 1900 he believed he had reached the end but the writing kept on coming and in May he sent off chapter thirty-one.

On 9 July, he announced that he had finished; he announced it again on 12 July, and then, on 14 July, he sent his wife, Jessie, and young son, Borys, to London before sitting down at 9 a.m. to write for twenty-one hours. Ink splattered across page after page, paper fell in piles to the floor, the room was a fug of smoke, the sun rose and then sank, and he put down his pen only when he had drained from Jim's jump the last drop of meaning. 'And that's the end,' Conrad wrote as he completed chapter forty-five. 'He passes away under a cloud, inscrutable at heart, forgotten, unforgiven, and excessively romantic.'

As Conrad gathered the pages together there came over the house a great silence. This is how he described the scene to John Galsworthy: 'Cigarette ends growing into a mound similar to a cairn over a dead hero. Moon rose over the barn, looked in at the window and climbed out of sight. Dawn broke, brightened. I put the lamp out and went on, with the morning breeze blowing the sheets of MS all over the room. Sun rose. I wrote the last word and went into the dining room. Six o'clock I shared a piece of cold chicken with Escamillo [the dog] (who was very miserable and in want of sympathy, having missed the child dreadfully all day). Felt very well, only sleepy; had a bath at seven and at 1.30 was on my way to London.'⁴ *Lord Jim* was published three months later to a mixture of astonished and exasperated reviews. Like Conrad's other novels and stories, it walked a tightrope between high modernism and light reading, the existential and the ripping yarn. It appealed to schoolboys who read nothing but the *Boy's Own Paper* and professors who placed it alongside their copies of Henry James.

Lord Jim can be told in a single sentence: Jim jumps from a sinking ship and then faces a life without honour. The tale was, Conrad wrote to his editor at *Blackwood's*, 'the development of *one* situation, only *one* really from beginning to end'. It is not a complicated plot and nor is Jim an enigmatic figure, but in Conrad's hands the story becomes manifold and Jim's consciousness a maze. 'Imagine a fat, furry spider with a green head and shining points for eyes, busily at work, some dewy morning, on a marvellous web,' wrote the reviewer for *The Critic*, 'and you have the plot of *Lord Jim*. It spins itself away, out of nothing, with sidetracks leading, apparently, nowhere, and cross tracks that start back and begin anew and end once more.'

Conrad's preferred method of narration is a story within a story: most of the book is contained in quotation marks, sometimes three in a row – ' " " – and it is Captain Marlow's role to hold the different frames together. *Lord Jim* opens with Conrad's narrator introducing Jim to the reader: 'He was an inch, perhaps two, under six feet, powerfully built, and he advanced straight at you.' In the fifth chapter, this first narrator disappears and Marlow takes over the story. 'After a good spread, two hundred feet above sea level, with a box of decent cigars handy, on a blessed evening of freshness and starlight,' Marlow tells a gathering of sea dogs how he attended the *Patna* inquiry and afterwards invited Jim to dinner in the Malabar Hotel.

At the end of chapter twenty-one, Marlow tells us that his 'last words about Jim shall be few', but his last words continue for another twenty-four chapters. *Lord Jim* is all talk: talk framed within talk; talk spinning itself away out of nothing. It is remarkable how much talk Jim's jump manages to generate. A column called 'Books to Cut' in the November 1900 issue of the British journal *Public Opinion* concluded: 'Words cannot describe the weary effect of all this . . . we long to get on and skip all this chatter, to discover into what sort of man Jim really develops.'

This is not the only occasion that Marlow talks about Jim. Throughout his life, Conrad tells us, 'many times, in distant parts of the world, Marlow showed himself willing to remember Jim, to remember him at length, in detail and audibly. Perhaps it would be after dinner, on a verandah draped in motionless foliage and crowned with flowers, in the deep dusk speckled by fiery cigar-ends.' When he remembers Jim, Marlow becomes 'very still, as though his spirit had winged its way back into the lapse of time and were speaking through his lips from the past'.

It is not only Marlow who talks about Jim. In Conrad's novel, wrote the reviewer for the *Daily News*, 'no one talks unless it is to discuss Jim'. The scandal had, Marlow says, 'an extraordinary power of defying the shortness of memories and the length of time: it seemed to live, with a sort of uncanny vitality, in the minds of men, on the tips of their tongues. I've had the questionable pleasure of meeting it often, years afterwards, thousands of miles away, emerging from the remotest possible talk.' Everyone has something to say about Jim, and much of it is extraordinary. One man, a German butterfly collector called Stein, compares Jim's psychological state to that of Hamlet, who also delays action and is then racked with existential doubt. '*Ja! Ja!* In general, adapting the words of your great poet: That is the question . . . how to be! *Ach!* How to be.'

On this particular night, two hundred feet above sea level, so much talk is generated by the facts of Jim's case that Conrad's reviewers mocked the novel's claims to realism. 'This after-dinner story, told without a break,' wrote Arnold Bennett in an unsigned review for *Academy*, in November 1900, 'consists of about 99,000 words. Now it is unreasonable to suppose that the narrator, who chose his words with care, spoke at a greater rate than 150 words a minute, which means that he was telling the after-dinner story to his companions for eleven solid hours.' Time is nothing to a sailor – this is the difference between shore people and sea people – and Conrad responded that 'men have been known, both in the tropics and in the temperate zone, to sit up half the night "swapping yarns" . . . whereas all that part of the book which is Marlow's narrative can be read through aloud, I should say, in less than three'. Other reviewers were baffled by the book. 'More readable novels, better novels in every way, have already been published by the score,' said *The Sketch,* but 'none more strange, none more genuinely extraordinary. *Lord Jim* is an impossible book – impossible in scheme, impossible in style. It is a short character sketch, written and rewritten to infinity, dissected into shreds, masticated into tastelessness. The story – the little story it contains – is told by an outsider, a tiresome, garrulous, philosophising bore. And yet it is undeniably the work of a man of genius.'

All this writing and rewriting, this torrent of words, came out of years of solitude and silence. '*Lord Jim* is a great book, a wonderful book, a magnificent book,' wrote William Alden (also, coincidentally, the name of Senator Smith) in the *New York Times Book Review* on 1 December 1900. 'Here, then, is a work of genius – of unique and superb genius . . . It is a phenomenon almost as strange as the author himself – the man who spent a lifetime at sea, dealing with the roughest places of life, and living wholly without books, then suddenly showing himself to be one of the most striking writers known to English Literature.' Perhaps the best description of the book's appeal came later from Albert Guerard, who wrote: '*Lord Jim* is a novel of intellectual and moral suspense, and the mystery to be solved, or conclusion to be reached, lives not in Jim but in ourselves. Can we, faced by the ambiguities and deceptions of life itself . . . apprehend the whole experience humanely? Can we come to recognise the full complexity of any simple case, and respond both sympathetically and morally to see Jim and his version of "how to be"?'[5]

Conrad, for whom English – which he learned aged twenty-one – was his third language, is never as relaxed as Marlow. This is one of the differences between the two captains. Captain Marlow is Conrad's English orator; his cadences are commanding and clear while Conrad's thick Polish accent weighed down a voice already heavy with gloom. Conrad was not a natural public speaker, but he was a conversationalist who knew the power of exchange. Talk for Conrad was romantic or it was nothing.

Edward Garnett, his first reader at Unwin's, describes the writer's conversation as 'a romance; free and swift, it implied, in ironical flashes, that though we hailed from different planets the same tastes animated us . . . there was a blend of caressing, almost feminine intimacy with masculine incisiveness'. When Marlow and Jim talk it becomes a sentimental education for them both: Jim's story is less interesting than Marlow's interest in Jim's story; that Marlow is fascinated by Jim makes Marlow himself fascinating, and Jim's words are rich because Marlow makes so much of them. More than any other writer, Conrad understands the difficulties of language: the subject on which he has most to say is the horror of being able to say nothing at all, of words drying up and failing the speaker. Speech might be no more a serried circle flying around an immovable fact, but so long as it keeps on coming, there is hope of some kind of meaning. For Conrad the modernist, meaning is always carved out of language but words are also 'the great foes of reality', as he puts it in *Under Western Eyes*. 'There comes a time when the world is but a place of many words and man appears a mere talking animal not much more wonderful than a parrot.'

When we first meet Marlow in *Youth* (1898), he is forty-two and drinking claret at a mahogany table with four other men who 'share the bond of the sea'. The narrator, one of the group, tells us that this story 'could have occurred nowhere but in England, where men and sea interpenetrate, so to speak'. Marlow – '(at least I think that is how he spelt his name)' – then relates the tale of his first voyage as a young sailor twenty-two years earlier. 'You fellows know,' Marlow begins, 'there are those voyages that seem ordered for the illustration of life, that might stand for a symbol of existence.' His ship, the *Judea,* is an ancient masted rig transporting 600 tons of coal to Bangkok. The coal spontaneously combusts and a fire breaks out in the bunker; after burning for days the ship is finally destroyed but the crew are saved. The fate of the *Judea* represents the end of the age of sail: all ships will soon be powered by coal. But the story is elegiac in other ways too: Marlow's subject is the exuberance of youth and its 'romance of illusions'. What to an older man is a harrowing experience at sea is, to a younger man, an awakening, an adventure. The days of his youth were when Marlow was happiest, and the group around the table drink to 'youth and the sea. Glamour and the sea! The good, strong sea, the salt, bitter sea, that could whisper to you, and roar at you and knock your breath out of you.'

We meet him for the second time in *Heart of Darkness* (1902), when Marlow is on board the *Nellie* telling his companions the story of Kurtz, an ivory trader who lives as a demi-god in the Congo. 'Do you see him?' Marlow asks; 'Do you see the story? Do you see anything?' The opacity of Conrad's prose sometimes makes it hard for his reader to see very much; E. M. Forster said that 'the secret casket of his genius contains a vapour rather than a jewel'. But

Conrad's aim as a writer is 'to make you hear, to make you feel . . . to make you see', to give the reader 'encouragement, consolation, fear, charm – all you demand – and, perhaps, also *that glimpse of truth for which you have forgotten to ask*'. From the introduction to *The Nigger of the 'Narcissus'*, this is the most perfect explanation in the language of why we read and also the most beautiful, the most extraordinary, of all Conrad's beautiful, extraordinary sentences. We read because we are looking for something we have forgotten, and this is why we need Marlow. He is a detective searching for clues.

Marlow wants us to see Kurtz and to see the *Judea* and to see Jim and to see the story; he wants us to see *everything*, even the invisible. But Conrad only wants us to see Marlow. Jim may be a coward or an idealist, a romantic or a criminal, but he is fixated on a single idea – the loss of his heroism. We see him as clearly, or as unclearly, as we see ourselves: the best of us would do what he did: Jim is Everyman, Marlow insists. But there are very few men like Marlow, men who live on land 'as a bird rests on the branch of a tree, so tense with the power of brusque flight into its true element that it is incomprehensible why it should sit still minute after minute'. Marlow is a detective but he is also a High Romantic; he belongs to the tradition of Romantic wanderers, like Coleridge's Ancient Mariner – which poem Conrad loved and whose lunar imagery suffuses *Lord Jim* – who have outlasted illusions but still thirst for something more. To Marlow, the quenching of this thirst can only be done at sea. Shore-dwellers are aliens whose solid world is closed, cluttered, imaginatively impoverished. Only in the presence of the endless monotony of the ocean can one immerse oneself in reflection.

It is typical of Marlow that he does not wander into *Lord Jim* until we have settled down into chapter four, at which point he kidnaps the story. 'That preposterous master mariner', as Henry James called him, is Conrad's other self, his double; Marlow is the English gentleman Conrad can never be. 'The man Marlow and I came together,' Conrad explained, 'in the casual manner of those health-resort acquaintances which sometimes ripen into friendships. This one has ripened. For all his assertiveness in matters of opinion [Marlow] is not an intrusive person. He haunts my hours of solitude, when, in silence, we lay our heads together in great comfort and harmony; but as we part at the end of a tale I am never sure that it may not be for the last time. Yet I don't think that either of us would care much to survive the other . . . Of all my people he's the one that has never been a vexation to my spirit. A most discreet, understanding man.'[6] Marlow 'lives as he dreams – alone' and therefore comes without a biography. A 'lanky, loose' figure with a 'narrow, veiled glance', Marlow is 'quietly composed in varied shades of brown robbed of every vestige of gloss'. He has no history, home, family or friends, no connections to the world in which he takes such an interest and of which he has such luminous understanding. He does not even possess a consistent personality; in Marlow the very idea of personality is obliterated. He is simply a mind in which 'some notion' is chased 'round and round . . . just for the fun of the thing'.

Like all mariners, Marlow is a threshold figure; wherever we find him, he is neither quite on land nor quite at sea. Whether on a deck or a verandah, in a port, a harbour hotel or moored at the mouth of a river, he is standing apart, looking on. Without having known the love of a mother, sister, wife or mistress, he is unencumbered by women. 'You say I don't know women,' Marlow explains in *Chance*, the novel in which he tries to get to know one. 'Maybe. It's just as well not to come too close to the shrine. But I have a clear notion of *woman*. In all of them, termagant, flirt, crank, washerwoman, blue-stocking, outcast and even in the ordinary fool . . . there is something left, if only a spark.' Sea creatures are less strange to Marlow than females. 'As to honour – you know – it's a fine medieval inheritance which women never got hold of . . . In addition they are devoid of decency. I mean masculine decency.' Over twelve years and in four separate books, Conrad needs Marlow to expound on women, youth, imperialism, cowardice, honour and fidelity. Only in *Chance*, where Marlow makes his final appearance, do we find him out of his element. This story of the blue-eyed, red-lipped Flora de Barral, suffocated by the combination of her husband's magnanimity and her father's incestuous love, was Conrad's least

satisfying and most commercially successful book. He called it his
'girl-novel'.

Conrad's writing is filled with irreversible acts, with men – and
occasionally women – who gamble and lose and are forced to live on.
When Marlow first talks to Flora de Barral, it is because he chances,
on a country walk, to prevent her from throwing herself from a preci-
pice. Always drawn to those who jump, Marlow then becomes
involved in Flora's decision to marry a sea captain whom she does not
love. 'The fact of having shouted her away from the edge of a preci-
pice, seemed somehow to have engaged my responsibility as to this
other leap.' The story is strung together by chance encounters, and
Marlow hears about the next stage of Flora's life when he happens
upon the second mate of the ship of which her husband was captain.
A jump, for Conrad, is just another sort of chance.

Ismay was thirty-seven when *Lord Jim* was published in 1900, but it is
unlikely that he knew the novel; it is a difficult read and he was not a
bookish man. He had probably never heard of Conrad, but Conrad
knew about Ismay. Everything about the *Titanic,* which Conrad
thought a 'monstrous' upholstered ferry – the excessive futility of its
conception, the excessive lack of professionalism amongst the management, the excessive number of words it was generating in the press
– offended his essential frugality. For centuries, travellers had been
blown about the sea on the movements of the wind like dispersed
dandelion parachutes and now they were shunted along by the screw-
ing motion of a propeller. The North Atlantic trade by which Ismay
made his living was, Conrad said in a letter to John Quinn after the
Titanic went down, 'not good enough for a man who cared for his
profession; very monotonous, very risky, no better than running a
tramway under disagreeable conditions of weather'.[7] Ships like those
run by the White Star Line, Conrad believed, destroyed the ancient
traditions of seamanship. 'A marvellous achievement', he said of steam
propulsion in his essay 'Ocean Travel', 'is not necessarily interesting.

It may render life more tame than perhaps it should be.' Sailing ships brought you closer to 'the silence of the universe', suspending you from the stresses and anxieties of land life. Under sail the traveller communed with the sea, but it was only by looking out of the curtained windows of the White Star liners that you remembered you were afloat at all. Conrad's ocean was not a house party but a place of monastic simplicity. As he puts it in *Chance*, 'the service of the sea and the service of the temple are both detached from the vanities and errors of a world which follows no severe rule'.

On 16 April, the day after the news of the *Titanic* reached England, Conrad offered a piece on the subject to *Nash's Magazine*. The editor turned his suggestion down and Conrad published a statement (now lost) in *The London Budget* on 20 April. Another article on the *Titanic* (also lost) appeared in *The Literary Digest* on 4 May. Needing to 'talk a little' about the wreck, 'for my own comfort partly', Conrad wrote to his agent on 22 April: 'In order to throw it off my chest I ask you to get Harrison on the telephone and ask him if he cares to get an article from me on the subject. A personal sort of pronouncement, thoughts, reminiscences and reflections inspired by the event with a suggestion or two.' Austin Harrison, the Harrovian editor of the *English Review*, was considering serialising *Chance*, which had, since January, been appearing in instalments in the *New York Herald*. Harrison turned down *Chance*, and accepted instead not one, but two personal pronouncements by Conrad.

The first of Conrad's articles, 'Some Reflections on the Loss of the *Titanic*', appeared in May 1912.[8] He noted the 'good press' the wreck had enjoyed, and condemned the 'provincial display of authority' exercised by the 'august' Senators Smith and Newlands, the 'grimly comic touch' they brought to the affair by 'rushing to New York and beginning to bully and badger the luckless "Yamsi" – on the very quay-side so to speak . . . What are they after? We know what happened. The ship scraped her side against a piece of ice, and sank

after floating for two hours and a half, taking a lot of people down
with her. What more can they find out from the unfair badgering of
the unhappy "Yamsi", or the ruffian abuse of the same?' The code-
name 'Yamsi', Conrad explains, is 'symbolic', used here to represent
not Ismay but commerce itself. While Conrad has no high regard for
shipping magnates, he must protest against the 'Bumble-like proceedings' of the Senate inquiry. What motivated such a tasteless rush to
abuse Ismay, 'a man no more guilty than others in this matter'? What
motivated the Senate to set up a court 'before the poor wretches
escaped from the jaws of death had time to draw breath', and before
the accusers themselves had time to learn the most basic sea terms?
The Senators did not even understand the language they were
required to speak. 'Such a simple expression as that one of the lookout men was stationed in the "eyes of the ship" was too much for
them.' Conrad could not see why there was an inquiry in New York
at all, the *Titanic* being a British-built ship which sank in high seas,
or why Ismay should answer the questions of 'any king, emperor,
autocrat, or senator of any foreign power'.

Conrad did not restrict his scorn to the Americans. He also
condemned the British Board of Trade, asleep on the job, which
'took its dear old bald head' out from under its wing to declare the
Titanic 'unsinkable' before putting it back again, 'in the hope of not
being disturbed for another ten years'. He condemned the ostenta-
tion of the *Titanic* herself, 'boomed' by advertising, ridiculous in her
Egyptian décor – or was it Louis Quinze? – with her 'gorgeously
fitted (but in chaste style) smoking room', her swimming pool and
'delightful French café', all of which gave the passengers a 'sense of
false security'. What is all this luxury *for*? Were there no ships, most
of us would happily put to sea in a bucket. The White Star Line are
dealers in illusion posing as 'benefactors of mankind' magnani-
mously 'engaging in some lofty and amazing enterprise'. And what
kind of discipline is operating on a ship in which passengers think
that entering a lifeboat is 'an optional matter'? What has happened to
the moral atmosphere of sea life, in which certain conditions and
rules prevail? 'The order to leave the ship should be an order of the

sternest character, to be obeyed unquestioningly . . . A commander should be able to hold his ship in the hollow of his hand.' Refusing to abandon ship when ordered to do so is evidence of social breakdown. The *Titanic* was not commanded, manned and equipped as a ship at all, she was a 'marine Ritz', a 'sort of hotel syndicate composed of the Chief Engineer, the Purser, and the Captain'. A 46,000-ton pleasure palace made of thin strips of steel was sent adrift to meet all the usual dangers of life on the waves, and there is uproar and 'surprised consternation' when she sinks.

His second article, 'Certain Aspects of the Admirable Inquiry into the Loss of the *Titanic*', which appeared in the *English Review* in July 1912, was written in reply to a letter he received from John Quinn.[9] 'The inquiry was a God-send,' Quinn had written to Conrad from New York, 'in that it lifted the cloud of mystery that shrouded the whole thing and was really a safety valve. It was a loosening by proxy of the pent-up horror that the loss had caused everywhere and by the time the inquiry was over people were more or less satisfied that they had got to the bottom of things . . . As a whole the American press behaved admirably both in news and editorial comments.'

The US Senate inquiry a God-send? A safety valve? It reached the bottom of things? 'The Senators of the Commission', Conrad responded through the pages of the *English Review,* 'had absolutely no knowledge and no practice to guide them in the conduct of such an investigation; and this fact gave an air of unreality to their zealous exertions.' And anyway, what is this obsession with bigness as a sign of progress? 'If it were, elephantiasis which causes a man's legs to become as large as tree trunks, would be a sort of progress, whereas it is nothing but a very ugly disease.' Bigness should not be confused with greatness; bigness has no intrinsic moral value, it is no more than 'mere exaggeration'. As for the language of heroism employed by the halfpenny press, 'there is nothing more heroic in being drowned very much against your will, off a holed, helpless big tank in which you bought your passage than in dying of colic caused by the imperfect salmon in the tin you bought from your grocer'. It would have been finer, Conrad suggests, 'if the band on the *Titanic* had been quietly saved, instead of being drowned while playing – whatever tune they were playing, the poor devils'. But Conrad is not, he insists, 'attacking' the shipowners themselves: 'I care neither more nor less for Lines, Companies, Combines and generally for Trade arrayed in purple and fine linen than the Trade cares for me . . . I am attacking foolish arrogance . . . I have been expecting from one or other of them all bearing the generic name of Yamsi, something, a sign of some sort, some sincere utterance, in the course of this Admirable Inquiry, of manly, of genuine compunction. In vain. All trade talk.'

Gone is Marlow's sinuous sidetracking, his serene suspension of moral judgement. It is Conrad speaking in these essays and his irony is out in full force. His tone is one of parental wariness, this is the 'I told you so' of the jaded and fastidious when the inevitable occurs. The *Titanic* was never a ship; she was a fashion. The focus on speed, on advertising, on profit, on the values of 'progressive' modernity which dominate the miserable affair offend Conrad's belief in the nobility of the sea, in the rigour and efficiency of those who try to combat it and the honour and decency of the men who devote them-selves to its traditions. The *Titanic* crew died, he believed, for commerce and it is his 'brother' seamen about whom Conrad is thinking when he writes about the disaster; their duty had at one time been his duty, their feelings were his feelings. As for the men who died 'compartmented' in the bowels of the ship, 'nothing can approach the horror of that fate except being buried alive in a cave, or in a mine, or in your family vault'. This 'horror' is as close as Conrad gets to commenting on the human side of the tragedy; he had no 'personal . . . thoughts, reminiscences and reflections' to make about the man at the tiller or the boys in the crow's nest, about the responsibility of the Captain, or the officers, or the role on the ship of the owner. He said nothing about the individual behaviour of anyone on the *Titanic*, and made it clear that he was using 'Yamsi' to refer to the corporation rather than to the man whose code-name it was. While the rest of the country celebrated individual acts of heroism and debated displays of cowardice, Conrad considered the *Titanic* to be no more than the tale of a tub. His essays are about an empty ship.

At the same time as he describes the *Titanic* as resembling 'a Huntley and Palmer biscuit tin', the romantic in Conrad understands, although he will never admit it, the romanticism of this particular ship, from its fatal name onwards. Conrad responds powerfully to symbols; he knows that the *Titanic* represents more than just profit, that there was something transgressive in her arrogant challenge to the gods, that she had about her an element of the sublime. Conrad says that he is on the side of the seamen but he is really on the side of the sea, that creature of 'unfathomable cruelty', and he both admires and mocks those with the audacity to float 'in the face of his frown'. The *Titanic* assumed that the ice would part for her, that she could control the waves, that her squash court and French café could dazzle and defy what for centuries has smashed boats and wrecked men simply because it could. The *Titanic*'s attempt to seduce the sea with her glamour appeals to the writer in Conrad.

The more distance he put between himself and his life as a mariner, the more Conrad idealised ships and the 'brotherhood' of sailors, and the more he romanticised the time in which he had been, for the best part, bored and disillusioned. He remembered the displacement of sail by steam as 'a swift doom' but the vessels had been operating side by side for years. Conrad constructed a myth in which an enchanted world abruptly ended and a commercial world began. 'No doubt,' he said, 'the days thus enchanted were empty, but they were not so tedious as people may imagine.' Or perhaps they were. What has changed for Conrad in the replacement of sail by steam is the role of the sea *in the imagination*. At one time, a traveller would leave the conditions of shore behind to find 'in the ship a new kind of home'. He now, especially when crossing the Atlantic, 'brings the conditions of shore life with him on board'. The sailing ship made a man free, while the modern steamer is a 'prison'.

Advertisement for the *Olympic* and *Titanic*, 1911

Conrad's problem is that, like Marlow, he is modern as well as being a Romantic. He is a product of the Titanic age with its love of technology and its 'speed lust', as a writer in the *Cornhill Magazine* described the acceleration of everything in 1909. Conrad had no interest in bigness, and none whatever in what E. M. Forster called, in *Howards End*, the 'clipped words' and 'formless sentences' of the new 'language of hurry'; but he, like Florence Ismay, loved mechanical velocity. One of the earliest motorists, he drove his Cadillac (which he pronounced 'Cadiyac') around the Kentish lanes at a belting pace, taking corners at thirty miles an hour, tooting his horn at foolish pedestrians, missing farm carts by shavings of inches. His young son Borys would accompany him, terrified, fascinated. Conrad in his motor car with his goggles and flat cap was like that other Edwardian speed merchant, Mr Toad: 'What dust clouds shall spring up behind me as I speed on my reckless way! What carts shall I fling carelessly into the ditch in the wake of my magnificent onset!'

Conrad's biographers describe him as living three lives – those of a Pole, a mariner, and an English writer – but his was really a double life. '*Homo-duplex*', he said, 'has in my case more than one meaning.' Conrad, as Virginia Woolf put it, 'was a compound of two men; together with the sea captain there dwelt that subtle, refined, and fastidious analyst whom he called Marlow'.[10] Marlow's temperament was closer to that of Conrad's father, the Polish nobleman, intellectual and leading Red activist, Apollo Korzeniowski, than it was to the nerve-racked, neurotic writer himself. Marlow and Apollo were both wanderers without land, except that Marlow relished his condition while Apollo, whose life was devoted to freeing Poland from Russian oppression, railed against it. Conrad described his father, whom many considered to have been a hero, as having an 'exalted and dreamy temperament; with a terrible gift of irony and of gloomy disposition'.

Józef Teodore Konrad Korzeniowski was born in the Ukraine in December 1857, five years before Ismay. To celebrate his birth, Apollo produced the following lines: 'To My Son Born in the 85th Year of Muscovite Oppression, A Song For the Day of His Christening'.

> *Baby son, tell yourself*
> *You are without land, without love,*
> *Without country, without people,*
> *While Poland, your mother, is entombed.*

Conrad's other mother, Ewa Bobrowska, was also entombed by the time he was seven, dying in northern Russia where Apollo was exiled in 1862. An only child, Conrad was raised by his heartbroken father whose stated object was 'to bring up Konradek not as a democrat, aristocrat, demagogue, republican, monarchist, nor as a servant and a flunkey of these parties – but only as a Pole'. Apollo was granted conditional parole in 1867 and he returned to Poland with his son, but his spirit and health were broken. The boy's companions were Apollo's 'clouded face' and his collection of Polish Romantic poetry. 'Poor child,' his father wrote, 'he does not know

what a contemporary playmate is.' Their life was one of confine-
ment; Apollo described the pair of them as the only two people 'left
on this earth'. Apollo's politics, he said, made him a 'monk in the
Polish order' whose thoughts were 'confined in the small cell of
patriotism' while Konradek was living 'as if in a cloister . . . we
tremble with cold, die of hunger, struggle in the abject poverty of
our brothers'.[11]

Apollo died in Cracow in 1869, his public funeral becoming a
mass patriotic demonstration; thousands turned out to pay a final
homage to Poland's great son whose tombstone reads: 'To the man
who loved his Homeland, laboured for it, and died for it – His
Compatriots. Apollo Korzeniowski – victim of Muscovite martyr-
dom.' Orphaned at twelve, Konradek was taken in by his
down-to-earth uncle, Thaddeus Bobrowski, whose letters to his
nephew provide the rare moments of levity in any Conrad
biography.

According to his Polish biographer, Zdzislaw Najder, Conrad
inherited from his country the 'heroic virtues of duty, fidelity, and
honour', while he inherited from his father a hatred of Russians, an
inflexibility of character, a preoccupation with public duty, a passion
for Romantic literature, an 'exceptionally intense emotional life' and
a tendency to pathological depression. It was Apollo who nurtured
the part of Conrad which, as Bertrand Russell put it, 'thought of
civilised and morally tolerable human life as a dangerous walk on a
thin crust of barely cooled lava which at any moment might break
and let the unwary sink into fiery depths'.

But Apollo did not pass down his love of the earth. His father's
obsession with a homeland left Conrad with only one option: to go
to sea, and aged seventeen he jumped . . . it seems. Bidding farewell
to Poland, where he would never live again, Conrad left for Marseilles
and the life of a sailor. 'I verily believe,' he reminisced, 'mine was the
only case of a boy of my nationality and antecedents taking a, so to
speak, standing jump out of his racial surroundings and associations.'[12] Conrad's writing is filled with boys who go to sea in order
to get away from home, but his critics and biographers puzzle over

his own desire to leave his landlocked country with its tragic history. He seems too cultured, too bookish, too depressive to endure the solitude and hardships of the nautical life. However, dreams of adventures in far-flung places are typical for young boys. 'What was truly extraordinary,' Najder reminds us, 'was his later career as a writer.'[13]

After two months in Marseilles, Conrad had his first experience aboard a ship, crossing to Martinique and Haiti on the *Mont Blanc*, a three-masted wooden barque. He went out as a passenger and returned as a member of the crew; from mixing with the Polish intelligentsia, he was now mucking in with French sailors. In his autobiographical writings, such as *The Mirror of the Sea*, Conrad would idealise these early years, picturing himself as the hero in a book, dabbling in gun-running along the Spanish coast, losing in love, feeling alive in his youth as never before or since. The tales of gun-running are quite possibly a fantasy and there is no evidence that Conrad was emotionally or spiritually transported by his early voyages. Instead, despite a healthy allowance from Bobrowski, he quickly fell into debt and despair and in 1878 he shot himself through the chest. The bullet, remarkably, went '*durch und durch*', narrowly missing the heart and not damaging any internal organ. Bobrowski, who rushed over from Poland to rescue his nephew, found him to be 'extremely sensitive, conceited, reserved and in addition excitable'. He sorted out his debt and put him back on the rails. Conrad's inability to handle his finances was to be a perennial problem for Bobrowski who, when asked again by him for money several months later, reminded the young man that 'you were idling for nearly a whole year, you fell into debt, you deliberately shot yourself – and as a result of it all, at the worst time of the year . . . and in spite of the most terrible rate of exchange – I hasten to you, pay, spend about 2,000 rubles'.

Several months after his suicide attempt, Conrad set foot in England for the first time when his steamer, the *Mavis*, landed in Lowestoft on 10 June. Already fluent in French, he began to learn English from the daily papers and 'from East Coast chaps, each built to last forever and coloured like a Christmas card'. He joined the British Merchant Navy and between 1880 and 1883 he sailed in Eastern waters, 'the place from which I have carried away into my life the greatest number of suggestions'. 'Karain', *Lord Jim, Youth* and 'The Secret Sharer' are all products of this period. In 1883 he took of a passage to Liverpool on a steamer; in 1886 he gained his Master Mariner's certificate and became a British subject, and in 1889 he went to the Congo from where he returned a different man. 'I am still plunged in the deepest night,' he wrote, 'and my dreams are only nightmares.' Still recovering from his breakdown, in 1891 he became first mate of the *Torrens*, one of the most famous clippers of the day, which ran a swift passage between London and Adelaide. The *Torrens* was, Conrad recalled in his essay, 'The Torrens: A Personal Tribute', 'a ship of brilliant qualities – the way [she] had of letting big seas slip under her did one's heart good to watch. It resembled so much an exhibition of intelligent grace and unerring skill that it could fascinate even the least seamanlike of our passengers.' The ship, with her cargo of 'a cow, a calf, and a good many sheep, geese, turkeys, ducks, hens, pigs . . . a kangaroo, two wallabies, five parrots, a dog, two cats, several canaries, and two laughing jackasses', was a veritable ark. The *Torrens* would be Conrad's only experience of working on a passenger ship and while he had a dislike of the general public, he now for the first time met educated Englishmen who spoke like Jim, using words like 'by Jove', 'ripping' and 'bally ass'.

In March 1893, the *Torrens* sailed from Adelaide with two new passengers. Ted Sanderson and John Galsworthy, both twenty-five, were returning from adventures in New Zealand, Australia and the South Seas, where Galsworthy had been sent by his parents in the hope that he might recover from what they considered to be an inappropriate love affair. Sanderson, accompanying his friend, was going home to Elstree where he was to help the Reverend Lancelot Sanderson run the family's prep school, while Galsworthy was reluctantly returning to his studies at the Admiralty bar. Both men had wanted to encounter Robert Louis Stevenson, now living in Samoa, whose novel *Treasure Island* they had read as teenagers. Published ten years earlier, *Treasure Island* told the story of another Jim who was faced with a jump:

'Jump', the Doctor orders Jim Hawkins. 'One jump, and we're out, and we'll run for it like antelopes.'

'No,' Jim replies. 'You know right well you wouldn't do the thing yourself – neither you nor squire nor captain: and no more will I.'

Jim Hawkins is one of the heroes Conrad's Lord Jim dreamed of becoming when, as a boy in his country parsonage, he was reading himself into ruin.

Galsworthy and Sanderson missed meeting Stevenson, but met Conrad instead. The Pole's 'rare personality attracted us at once', Ted Sanderson wrote, 'and a friendship was begun which lasted unbroken until death'. For Galsworthy, the experience of talking to Conrad 'outweighed . . . all the other experiences of that voyage'. He remembered Conrad as 'tanned, with a peaked brown beard, almost black hair, and dark brown eyes, over which the lids were deeply folded. He was thin, not tall, his arms very long, his shoulders broad, his head set rather forward. He spoke to me with a strong foreign accent. He seemed to me strange on an English ship. For fifty-six days I sailed in his company . . . Many evening watches in fine weather we spent on the poop. Ever the great teller of a tale, he had already nearly twenty years of tales to tell.' During the evening watches, Conrad told tales of his own 'romantic history': Poland, the South Seas, gun-running in Spain. He was, Ted Sanderson said, 'a fascinating talker on almost any subject'. Conrad's great characteristic, Galsworthy concluded, was 'fascination'.[14]

Neither Conrad nor Galsworthy had yet become writers, but the manuscript of Conrad's first novel, *Almayer's Folly*, currently in the locker of his cabin on the *Torrens*, would be finished later that year at Elstree School. Here Conrad stayed for ten days as a guest of the Sanderson family, who immediately took to him. Elstree was Conrad's 'dear place'. 'I have been made at home in Elstree,' Conrad wrote to Agnes, one of Ted's sisters, 'by so much kindness; I am often there in my thoughts.'[15] The Sanderson household gave Conrad his first experience of family life, and he later dedicated *The Mirror of the*

Sea to Katherine Sanderson, Ismay's Mrs Kitty; his second novel, *An Outcast of the Islands*, about a man on the run from a scandal, he dedicated to Ted Sanderson. After Conrad married and started a family of his own, Sanderson promised his friend's five-year-old son, John, a place at Elstree, but the boy went instead to Ferox Hall at Tonbridge, run by Ted Sanderson's sister, Agnes. So by curious coincidence, the spot in which Ismay had been least happy, Conrad was most happy; the place where Ismay had first felt homesick, Conrad first felt truly at home.

This voyage on the *Torrens* was to be Conrad's last. In no other kind of life, he said in *Lord Jim,* than that of the sailor 'is the illusion more wide of reality – in no other is the beginning *all* illusion – the disenchantment more swift – the subjugation more complete'. Aged thirty-seven and already an exile, he made another standing jump, leaving his world of ships for life at a desk. It was now that Konradek transformed himself into Joseph Conrad, and began the process of extracting from his years at sea every drop of meaning, every ounce of significance, as though meaning and significance were what he feared these years may have lacked.

6

The Secret Sharer

A man always has two reasons for the things he does: a good reason
– and the real reason.

J. Pierpont Morgan

A man's most open actions have a secret side to them.

Joseph Conrad, *Under Western Eyes*

Florence Ismay and the children were on a motoring holiday in
Wales when they heard that the *Titanic* had sunk. They returned to
Liverpool straight away and on Wednesday 17 April, Florence wrote
her husband a letter.

My darling Bruce,
   It seems an eternity since last Monday night when I first heard
that the *Titanic* had met with an accident. We reached Fishguard
about 6.30 and there got telegrams from the Office and from Margaret
[their eldest daughter] saying that there was a report that the ship had
struck an iceberg, but no lives lost, and that she was proceeding with
two other liners to Halifax. Of course I was full of sorrow at the
thought of the splendid vessel but felt no real anxiety until the next
wire about 10.30. That was followed by another at 3.30 which made
me terrified for your safety. Words fail to describe the horror and
anxiety of the hours that followed. It seemed an eternity.

Oh darling, what that time must have been like to you. When I think of the anguish you must have been through it makes me tremble even now. Thank God. Thank God that you have been spared to us. My life would have been over if you had not been saved. For me there never has been and never could be any man but you and I feel I can never express the gratitude and thankfulness that fills me for your escape . . . I have wished many times since Monday night that I had gone with you, I might have helped you in this awful hour. I know so well what bitterness of spirit you must be feeling for the loss of so many precious lives and that thing itself that you loved like a living thing. It must have been ordained by Providence as every precaution human skill and care could devise had been done. We have both been spared to each other, let us try to make our lives of use in the world. My dearest, if I have you I feel no trouble or sorrow can be unbearable, and that these last 48 hours that we can never forget may yet in some unknown way be turned into good . . .[1]

This is possibly the first love letter that Florence has ever sent, and she reveals her feelings for her husband as though they are a secret she has harboured for years. She can hardly bear what he has been through; it is pure providence that he is still alive, that their marriage has been given a second chance. Bruce, who has never before shared anything with his wife, will now need her to help carry his load. She has forgiven him the errors of the past; they will start their lives together again and this time do it differently; they will make themselves of use to the world and approach each day with gratitude and generosity. Most of all, they will now talk to one another, the decades of silence are over. Bruce is due to retire soon anyway: he will leave the sea for good, they will no longer need to be apart. And this might indeed have been their story, had Ismay not found a secret sharer in Mrs Thayer.

The *Titanic*'s maiden voyage has always conjured up images of ill-fated lovers. Whether it is the ship and the iceberg in Thomas Hardy's poem, or Jack and Rose in James Cameron's film, the story is one of convergence. When Ismay and Marian Thayer first encountered one another we don't know; they may have met through friends in New York but it is more likely that they met at sea and consolidated their relationship on board the *Titanic*. Ismay was at his most relaxed on his ships and the Thayers frequently crossed the Atlantic on White Star liners. In his account of the wreck, Marian's son, Jack Thayer, said that he and his father spent a good deal of time talking to Ismay on the *Titanic*, that during the voyage they got to know him 'well'. Friendships which would never form on land blossom on board ships where the usual bounds and restrictions are suspended and passengers and crew exist in a zone between time and place. Conrad's talk on board the *Torrens* 'fascinated' Galsworthy and Sanderson; his 'rare personality attracted us at once'; Ismay's attraction to Mrs Thayer's talk and rare personality was also immediate. She was the most sympathetic woman he had ever met.

Marian Thayer belonged, like the Astors and the Guggenheims, to Scott Fitzgerald's enchanted world of East Coast money. From impeccable Philadelphia stock, she had married John Borland Thayer, ten years her senior, on her twentieth birthday. Before taking on the Vice-Presidency of the Philadelphia Railroad, Thayer had been one of the state's premier cricket players. Three months before the *Titanic* set sail, Mrs Thayer's presence at Philadelphia's First Assembly Ball had drawn the usual press attention. This was 'the Prime Social Event of the Season' and the lovely Marian Thayer was 'one of the most strikingly costumed women' of the night. She wore, according to *Women's Wear Daily*, a high-waisted white satin gown with long train and a 'gauzy overdress, heavily spangled with gold'. Her slippers were also gold, and her thick dark hair shone in coils around her head. In love with her husband, devoted to her four children – the youngest of whom, Pauline, was eleven when the *Titanic* went down – and loyal to her friends, Marian Thayer was the image of contentment but seemed also to understand unhappiness, and this was her appeal. Her husband may have suffered from bouts of melancholia; he was periodically described in gossip columns during 1910 and 1911 as being unwell and in need of rest, and the Thayers visited 'nerve doctors' in Switzerland. It was

typical of Mrs Thayer to have insisted on the grief-stricken Emily
Ryerson's taking a walk with her on the deck of the *Titanic* the
afternoon that they met Ismay; the grieving and the troubled gravitated towards her and she drew them in. Graceful and low-voiced,
she shirked the superficiality of groups, preferring the intimacy of
one-to-one conversation. She was a collector of life stories; people
talked to Mrs Thayer as they would to no one else. She soothed the
lonely with her promise of friendship, she reassured the bereaved
with her belief in an afterlife. There was no pain that Mrs Thayer
apparently could not take away.

The night the *Titanic* went down, the Thayers had been at the
dinner for the Captain hosted by their friends George and Eleanor
Widener. Other guests included William Carter, who jumped into
the same lifeboat as Ismay, Carter's wife, Lucille, who would divorce
him on account of it, and Major Archie Butt, the trusted aide of
President William Howard Taft. During the evening, while Ismay
was dining with the ship's doctor on the other side of the room, Mrs
Thayer locked herself in private talk with Major Butt, who was
returning on the *Titanic* from a European holiday which the President
had advised his 'dear Archie' to take.

Six days later she described their conversation in a letter she felt
'compelled to write' to the President. 'In my own grief', Mrs Thayer
began, 'I often think of yours.' She addresses the most powerful
man in America as a fellow suffering being. 'I feel I must write to
tell you how I spent the last Sunday evening with Major Butt – for
we all cherish news of last hours – and we spoke much of you. How
devoted to you he was, and oh what a lovely noble man he was!'
Her meeting with Major Butt was '*meant* to be', Mrs Thayer says;
he 'opened his heart to me and it was as though we had known each
other well for years'. This was the first time that she had ever spoken
to Major Butt, but 'from the moment we met [we] never turned
from each other for the rest of the evening'. She felt 'as though a
veil was blown aside for those few hours illuminating distance
between two who had known each other always *well* long long
before and had just found each other again – and I believe it'.

While the rest of the room discussed the stock exchange and cross-ing speeds, the society wife and the forty-six-year-old bachelor were 'opening our innermost thoughts to each other'. He spoke of his love for his mother ('he told me I was just like her') and his sister-in-law, Clara, to whom he wrote every day about his life in the White House with Presidents Roosevelt and Taft, letters which were later published. He spoke of his 'love' for the President and of 'someone else he loved but that I do not'. They made an engage-ment to meet again the next day, 'as I was going to teach him a method of control of the nerves through which I had just been with a Swiss doctor, knowing it would be a very wonderful thing for him if he could get hold of it for he was very nervous and did not know how he was going to stand in the rushing life he was returning to, and we were going to work so hard over it the rest of the time on board'. She ends by asking if the President thinks there is 'any *chance* of seeing my husband or [Major Butt] again here in this life? My reason tells me No but how can we give up all hope until some days yet go past of this cruel torture?'[2]

Marian Thayer's appeal is disarming. She has a belief system, she is searching for a deeper truth; her interests are in fate, the spirit world, alternative medicine; she is intense and odd and immediate. She sanctifies human relations; she describes her conversation with Major Butt as though it were the beginnings of an affair, she writes a letter of condolence as though it were a love letter. She speaks to everyone in the same way. Her genius lies in the realisation that Taft responded to the *Titanic* not as an international catastrophe but as a personal loss. 'I cannot turn around in my room,' he wept at Butt's memorial service; 'I can't go anywhere without expecting to see his smiling face or hear his cheerful voice in greeting.' While the former President, Theodore Roosevelt, who had also been close to Butt, was sending his 'deepest sympathy to the kinsfolk of those who have perished', Taft was thinking of the death of one man only: 'Archie was the soul of honour. He was wholehearted and wholesome; courteous and courageous and a charming compan-ion.' While others thought Taft's self-involved grief inappropriate at a time when he was required to steer his own ship through rocky waters, Marian Thayer opened her arms to him. 'Never had I come in such contact *immediately* with anyone,' she told Taft of her encounter with the Major. 'He felt the same and we both marvelled at the time at the strangeness of such a thing . . .' But there was a quality of strangeness to all her relationships, particularly the one she shared with Ismay.

———

It seems likely that while they were on the *Titanic*, Mrs Thayer came into 'contact *immediately*' with Ismay in the same way she had done with Major Butt. Ismay, who was not used to having 'contact' with anyone – apprehension was the response he more usually generated in people – interpreted their easy communication as a mutual attraction. She flattered him with her attention as she probably flattered other men; women were raised to reflect men at twice their size, and in response Ismay felt unusually relaxed and open. Conrad noted in *Victory* that forty-five is the 'age of recklessness for many men'; in the American edition of the novel he increases it to fifty. During his time on the *Titanic*, Ismay experienced the recklessness that comes from supreme self-confidence and sudden, unexpected joy. His ship was the future; he was enjoying the attention of a beautiful female passenger. Having tried to get out of being present on the voyage, he was now in no hurry to get to New York. Finding time alone with Marian Thayer, who was constantly chaperoned, was impossible but when they met in company, whether in the dining room or on the deck, they never turned from each other. He looked for reasons to detain her, and when he saw her on the promenade deck with Emily Ryerson on the afternoon of Sunday 14 April, he had the perfect excuse: he showed her the message from the *Baltic*. While Mrs Ryerson told the story to the press, Mrs Thayer said nothing about the encounter herself. When the time came for Mrs Ryerson to sign her affidavit, it may have been Marian Thayer who persuaded her to withhold

Ismay's comments about lighting further boilers to get through the ice. The poor man was going through enough already.

Mrs Thayer phoned Ismay to say goodbye before he left New York on 5 May, and he sent a message to her from the pilot boat as he was leaving on the *Adriatic*. A letter from Mrs Thayer was then waiting for him when he boarded the ship which, he said, he 'read every night'. Her letter eclipsed the one from Florence, which he received several days earlier: there is nothing more repellent than a love letter from a person you no longer love. Marian Thayer had lost her husband but her son, Jack, who had also gone down with the ship, had been miraculously saved by clinging on to the same upturned lifeboat as Lightoller. She at least had something to be thankful for and, unlike the other widows in first-class, harboured no ill will towards Ismay. Instead, she gravitated further towards him. Such was Mrs Thayer's sympathy for Ismay that while the other widows on the *Carpathia* were spreading rumours about his culpability, Jack went to see him in his cabin. 'I have never seen a man so completely wrecked,' he recalled twenty-eight years later.[3] Jack Thayer's description, which he must have taken straight to Marian, suggests that he saw Ismay not as his father's killer but as his mother's friend. To Mrs Thayer, Ismay was neither a villain nor a coward but someone who was grieving, as was she. Her sympathy was a cause of irritation to the other widows on the *Carpathia*, particularly Eleanor Widener, whose dead husband's father was on the board of the IMM, and who had not liked Ismay sufficiently to invite him to their dinner party. 'Better a thousand times a dead John B. Thayer than a living Ismay,' Mrs Widener announced when she heard that the president of the IMM was alive. Her remark was a reminder to Marian Thayer not to let a soft heart besmirch the memory of a heroic husband.

In the letter to Ismay which she sent to the *Adriatic*, Mrs Thayer said that she wished she too had died that night, that death would be better than living like this, and she asked him not to forget her.[4] She enclosed a page cut from her calendar which carried some words which had given her comfort and which she hoped would comfort him as well. Ismay replied from the ship that he 'repeated the words regularly' and that he would keep them on his dressing table. 'Ever since leaving America,' he continued, 'you have been in my thoughts and I have talked to you so much. How I wish you had been here as I am sure we could have helped one another to bear our grief and loss. For you, poor dear, my heart bleeds and I cannot convey to you in words all I feel. I ask myself often why it is you seem so much to me. Can you answer?'

His crossing on the *Adriatic*, he told her, has so far been 'wonderful'. While he stayed in his room for the first few days, he now started going onto the deck twice a day for a short time. 'I hate seeing anyone I do not know, as I cannot feel I have done anything wrong and cannot blame myself for the disaster. Still, the feeling is there. I can only hope time will make things easier.' Ismay had confided to other people his sense of blamelessness. In the letter written to the brother of Frank Millet on 25 April, he had also expressed his belief that 'I am not in any way responsible for the truly dreadful disaster'. The death of Mrs Thayer's husband, Ismay now tells her, had nothing to do with him; it was an act of God. But still, he says, 'the feeling is there'. The feeling of responsibility has attached itself to his skin like an irritating tick which, with luck, he will be able to flick away. 'Forgive me for inflicting on you such a long letter,' he ends. 'Goodbye, my friend.'

There are few people less equipped than Ismay for a journey into the human heart but in Mrs Thayer, it seems, he has found his Marlow: 'I would like somebody to understand – somebody – one person at least!' Jim reveals in the Malabar Hotel, 'You! Why not you!' Ismay has never known the comforts of fellowship, but Marian Thayer talks intensively to many people. She is surrounded by family and friends and she spends her days writing to those who share her sense of the meaninglessness of living on after such a catastrophe; she writes to the President, she writes to Ismay, she writes to the other American first-class wives who lost their husbands, many of whom are friends of hers, some of whom are neighbours. She will stay in touch with Captain Rostron for the next ten years. Virginia Woolf believed that 'there should be threads floating in the air, which would merely have to be taken hold of, in order to talk. You would walk about the world like a spider in the middle of a web.' Mrs Thayer worked her own loom of human relations. In this, as in other things, she flowed in the current of the age.

The Edwardians 'deified', as E. M. Forster put it, 'personal relations and expected them to function outside their appropriate sphere'. Human beings, it was felt, formed a part of one another; personal connections were occasions for epiphanies. Mrs Thayer experienced an epiphany when she talked to Major Butt, Ismay had an epiphany when he talked to Mrs Thayer. Lucy Honeychurch and George Emerson experience their first embrace in *A Room with a View* (1908) as a divine revelation; in that moment they 'crossed', as Forster says, 'some spiritual boundary'. In *Howards End* (1910), Forster reiterates the mantra of the day, 'Only Connect': it is 'personal intercourse, and that alone, that ever hints at a personality beyond our daily vision', this is what will win us 'immortality'. For Mrs Thayer, who had long tried to make sense of life and death by finding patterns of order, cohesion, balance and oneness, human relations had become an answer for all of this horror. She believed that she and Major Butt had met in a previous life; she believed that she had crossed over from the other side in the past and that the lost souls of the *Titanic* may therefore cross too. She was deeply interested in spiritualism. No doubt, while she was on the *Titanic* she also talked to the journalist and social reformer W. T. Stead, founder of the spiritualist journal *Borderlands*. Stead, who predicted during séances his own death by drowning, had published an article in 1886 called 'How the Mail Steamer Went Down in the Mid Atlantic by a Survivor'. 'This is exactly what might take place,' he concluded, 'if liners are sent to sea short of lifeboats.'

'Writing to you,' Ismay now tells Mrs Thayer, 'makes me feel as if I was speaking to you. How I wish I was.' When he writes that since leaving America he has 'talked to you so much' he is adopting

her particular language: she believes that communication is not
dependent on presence. 'You are', Ismay repeats, 'constantly in my
thoughts'; forgetting Mrs Thayer would be 'absolutely impossible'.
They are the same in their 'grief and loss', she for her husband, he
for his ship. They are both missing the centre of their lives, neither
of them now has a future. But occasionally Ismay feels a flash of
hope: he does perhaps have a future – Marian Thayer, in some
form or other, is his future. His relief is so overwhelming that it
feels like love. She understands so much – her great gift is her
understanding, she is practically telepathic – but can she understand something as strange as his feelings for her? Ismay knows
that she can. He is sure she is 'acting bravely and not giving way'
to her desire to give up her life, that she is 'thinking of others who
are so dependent on you and who you love'. He is 'absolutely
convinced that we will meet again very soon. Something tells me
this and I am satisfied we will always be a good deal to each other.
You know what I mean.' Marian Thayer knows what everything
means, this is what is so overwhelming for a man like Ismay, who
finds expression difficult. He does not need to spell anything out.
But equally wonderful for Ismay is that *he* also knows what he
means: for the last three weeks he has not known what anything
means. What Conrad describes himself as doing for Augustine
Podmore Williams, his model for Jim, Mrs Thayer does for Ismay:
she 'seeks fit words for his meaning'. 'You must not be morbid or
distressed,' Ismay tells her. 'As you say, the easiest way would be to
join those who have gone before. I well know the feeling. As I've
told you, time alone is the healer of our great sorrows and we must
thank God that this is so.' Ismay is trying his best to carry on Slow
Ahead.

———————

The *Adriatic* landed in Queenstown on 10 May and Florence went on
board to meet her husband, about whom she had not stopped thinking for nearly a month. She now knew that he was not the hero of the

hour, that he had jumped into a lifeboat with the women and children, but she was still grateful. So many wives were without their
husbands, so many children without their fathers, but her family had
been saved from such a tragedy. She was desperately upset, particularly at Ismay's treatment in America, but also curiously elated; the
chance of his survival, and the chance they, as a couple, had been
given were both so extraordinary.

Ismay was the last passenger to leave the *Adriatic,* and the *Daily
Sketch* reported that although he 'seemed to be suffering from
nervous strain when he embarked . . . his health has improved
during the voyage'. The next day he was back in Liverpool, where
he was met by cheering crowds waving hats and handkerchiefs.
Once more looking 'pale and haggard' he acknowledged, through a
spokesman, the kind messages he had received, 'which he very
much appreciates in the greatest trial of his life'. It was unclear to
the press which trial he was referring to: the wreck of the *Titanic* or
his treatment in the US inquiry. The spokesman also requested that
Ismay not be pestered for a statement, 'first because he is still suffering from the very great strain of the *Titanic* disaster and subsequent
events; again, because he gave before the American Commission a
plain and unvarnished statement of facts which has been fully
reported, and also because his evidence before the British Court of
Inquiry should not be anticipated'. The 'statement of facts' given
by Ismay to the US inquiry had not been 'plain and unvarnished',
as anyone who had followed the proceedings in the British papers
would know, and the reference to the 'round unvarnished tale' by
which Othello woos Desdemona must have been Ismay's own. 'She
loved me for the dangers I had passed,' Othello explained to those
who could not understand how the fair Desdemona could feel so
'unnatural' a love, 'and I loved her that she did pity them'. Ismay
was thinking of Marian Thayer.

He returned to letters of condolence, but also anonymous letters
and – most terrible of all – letters from those wanting comfort themselves. His world had been reduced to endless words on paper. On 15
May, Ismay opened the following:

Sir,

I am writing to you in reference to my dear brother Arthur Hayter who was a steward on the ill-fated *Titanic* and was among the drowned. I heard from his wife that he had you to look after during the voyage and I thought perhaps you might be able to let us know if you saw anything of him at the last, on that fatal night. It would be a little consolation to us, his broken-hearted brothers and sisters, and to his aged parents who are 82 and 84 respectively. He was such a good brother and son. Excuse me for taking the liberty of writing to you, and thanking you, sir, for a reply.

I remain, yours sincerely,

Louise Hayter[5]

Ismay's reply was doubtless along the same lines as his letter from Washington to the grieving brother of Francis Millet. He heard also from Lucille Carter, whose husband had shared his life-boat. Mrs Carter was glad to hear that Ismay had been welcomed home, and appalled at the way he had been treated by the American press. William E. Carter's own survival had been of no interest to the papers, nor had he been asked to give evidence at the US inquiry.

Ismay was home, but sea-changed. His nightmares woke the house, he was blackballed from his club, an old friend turned him away from the front door. Florence realised that their lives, as she put it, were 'ruined'. Ismay was the loneliest man in the world, and at the heart of his aloneness lay a horror she could do nothing to soothe. Florence believed, as Bruce did too, that he had been saved by the will of God, but the 'feeling was there' for Ismay that it was his own will which had saved him, that his survival, rather than being divine intervention, had gone against the natural course of things. In build-ing the *Titanic* he had placed his trust in the material rather than the spiritual world, and God had now abandoned him. The air was thick with theories of a universe determined by human rather than divine strength. 'I am the master of my fate,' wrote Conrad's admirer, W. E. Henley, 'I am the Captain of my soul.' It was hard for Ismay to hold on to the belief that it had been God's wish that he jump into Collapsible C, or that his life had been saved for any purpose other than to experience hell on earth.

He did not want his wife's easy sympathy; he was appalled by her suggestion that the disaster would bring them closer together. However hard she tried, Florence, he believed, could never understand what he had experienced because she had not been there. Their marriage had changed after the death of their eldest son in 1891 – neither had recovered from the shock and grief – and here was Marian Thayer offering to console him over the death of 1,500 people. Florence now made a decision she would later regret: the *Titanic* was never to be mentioned again in her husband's presence. He was not to be allowed or encouraged to talk about the ship, the whole thing was to be forgotten. In addition, Florence decided that Bruce had behaved impeccably. She was not going to face the world as the wife of a coward. As Conrad put it in *Victory*, 'the last thing a woman will consent to discover in a man who she loves, or on whom she simply depends, is want of courage'.

———

Mrs Thayer replied to Ismay's first letter with a long and 'splendid' one of her own; so long indeed that she feared he would be disinclined to read it all the way through. Everyone is being very kind to her, she says, but she thinks she will have to give up their wonderful house, and she doubts that she and Ismay will ever meet again. Ismay, who read her letter through many times, waited until Florence had gone out before settling down to compose his reply. 'My wife has gone to church and I am sitting writing to you by the open window, looking over the garden. Oh, how I wish you were here and we could sit out in the garden and help each other. It would be lovely. I feel, now, that you are very close to me. I wonder if you are.' Ismay was desperate to talk about the *Titanic*; he fell upon Mrs Thayer as he fell upon the journalists at the Waldorf-Astoria when he felt misrepresented by Senator Smith. For a man whose 'watertight compartments', as Forster said of Lucy Honeychurch in *A Room with a View*, 'never broke down', Ismay often finds it hard to contain himself. 'How I wish I were near you,' he says to Mrs Thayer again; 'I know I would never be angry or vexed by anything you said.' It is wonderful, he feels, that he can open up to her like this. 'Really at times the outlook for the future is very very black,' he continues, 'and I wonder if I can face it. Everyone here is most kind and sympathetic but I feel my heart and spirit are broken and feel at times I must give in. But why mention my troubles.' He is sending her some lines which he tells her to put on her own dressing table:

> *With cheerful steps the path of duty run*
> *God nothing does, nor suffers to be done,*
> *But thou thyself wouldst do it, could thou see*
> *The end of all events as well as he.*

Engraved on the silver frame are Ismay's initials and Marian Thayer's, and the date the *Titanic* went down. Ismay is 'absolutely certain' that they will meet again, 'and perhaps sooner than you think'. The most 'difficult thing', he explains, is 'taking some interest in [my] surroundings . . . I cannot interest myself in anything. My mind seems to refuse to retain anything but the recollection of that awful disaster.' Time will help, he continues to hope. On lighter matters he wonders if she has seen Philip Franklin lately – 'What a truly splendid man he is – loyal and true' – and whether young Jack – 'a boy to be proud of' – has now gone to college? His letter is becoming too long, he will be boring her, 'but you must forgive me as I feel it is the next best thing to talking to you'.

On 31 May, Marian Thayer and her relation, Florence Cumings – who also lost her husband on the *Titanic* – go to a luncheon given by Madeleine Astor, the teenage widow of John Jacob Astor. The guests of honour are Captain Rostron and Dr McGhee of the *Carpathia*, but it is unlikely that Mrs Thayer shares with either man her knowledge of Ismay's current state of mind. She gives the doctor a gold cigarette case and Mrs Astor presents Rostron with a gold watch. The following day, the Captain and the doctor accompany Mrs Thayer back to her home in Philadelphia, where she holds a dinner for them and includes amongst the guests several other *Titanic* widows. Mrs Thayer belongs to a community of grief.

In London, the British Board of Trade inquiry into the wreck has been in progress since 2 May, but Ismay does not make his first appearance on the witness stand until 5 June. He follows the proceedings in the press, but does not mention the inquiry to Mrs Thayer until his ordeal is over. His next letter is written on 18 June, when he is staying at his brother's Scottish hunting lodge. 'Of course I could not misunderstand anything you write,' he reassures her. 'You must never think of such a thing and I hope you will always write me exactly as you feel.' He hopes she is taking an 'interest in matters' and admits he has 'been very depressed lately'. He has 'lost all desire for living' and sees 'no future' for himself. He says that he is sorry that she was 'asked to give evidence' to the Senate inquiry and regrets that 'it must have been very trying'. In her affidavit, Mrs Thayer had described 'rowing continuously for nearly five hours'.

Ismay has been in Scotland for ten days now. 'It is the wildest place you can imagine and you never see a soul. I spend my time fishing and walking and am not tempted to do any work. I could not do so if I wished to. The doctor tells me the strain I have been through has affected my heart and it will be a long time until it recovers. Any ambitions I had are entirely gone and my life's work is ruined.' He speaks of the loss of his future ambitions, but he had no future ambitions; Ismay had planned to retire the following year. He tells Mrs Thayer that 'I can never again take any interest in business and I never want to see a ship again, and I loved them so and took such an interest in the captains and the officers. What an ending to my life. Perhaps I was too proud of the ships and this is my punishment. If so it is a heavy one.' He is sinking into himself; he has reduced the enormity of the disaster to a personal tragedy, a blight on the quality of his life. The only other life it has affected is Mrs Thayer's, but her pain is nothing compared to his. She is not hiding out in a hunting lodge in Scotland; she can at least show her lovely sad face to the world; she is not being blamed for the whole thing. He will send her a photograph of himself as she requests, but he dislikes being

SANDHEYS.
MOSSLEY HILL.
LIVERPOOL.

Dearest Father;
This is just a line to let you
know how sorry I am that I did not
see more of you today, and to tell you
that I quite realize what an ordeal
you have had to go through and
how deeply I feel for you however
I very much hope that the worst
is over now and that you will
never again be misjudged and
your words misinterpreted as
they have been in the present
inquiry, I hope you will be
benifited by your stay at Dalnas-
'pidal and not be worried by
and anonymous comunication.

photographed and has only one picture which was taken a few years ago. She is in his 'thoughts' and he invites her to 'come to see us' when she is next in England; 'It would be such a pleasure'. He closes by apologising for having 'bored' her with 'such a long letter but it is the next best thing to talking to you', and repeats that he is certain they will meet again.

During his stay in Scotland, Ismay receives a nervous letter from his son Tom, who at seventeen is the same age as Jack Thayer. 'I very much hope that the worst is over now, and that you will never again be misjudged and your words misinterpreted,' Tom writes to the man who rejected him when he was left crippled by polio. He also hopes that while his father is in Scotland, he will 'not be recognised' or 'worried by any anonymous communications . . . I know this letter is very badly expressed, but I hope you will realise the spirit in which it is written is none the less sincere for that.'

Ismay now hears about an isolated house by the sea in Galway, on the west coast of Ireland. It is wild, has superb salmon fishing, and is on the market. He buys it, sight unseen: Ireland, where the *Titanic* was born, will be his future.

———

Ismay wanted to make a second statement to the press in which he could explain, in his own words, his behaviour on the *Titanic*. 'I don't think the portion of the public whose opinion we value,' Harold Sanderson carefully advised him, 'requires any further instruction as to your views or the merits of your action. The subject is dying and no longer interests them and I would leave it alone.' Prevented from talking about the disaster both in private and now in public, Ismay told Sanderson that 'I am afraid we look at the position from entirely different points of view; you have not been attacked, whereas I have, so you can easily afford to sit still and do nothing.'[6] Ismay had never been good at doing nothing.

The report of the British Board of Trade inquiry was published on 30 July, after which Franklin came over to England on White Star business. He spent some time with Ismay and returned to New York on the *Olympic*, where he wrote thanking him for his 'frank and generous hospitality . . . you have always been most courteous and kind to me, and it was always a great pleasure and honour for me to serve under you'. Because he has never before travelled on the *Olympic*, Franklin 'had no idea what a splendid, comfortable and marvellous steamer she is, but all the time I cannot help think-ing of your good self – the man that had the nerve and ability to

order and plan her, and then of what happened, and it all seems too
cruel, but I suppose there was some good reason for it'. Franklin,
who has only now been told that Harold Sanderson is to take over
the presidency of the IMM, 'feels hurt' that no one informed him
of the secret arrangement. He understands that 'recent develop-
ments' have made it 'clear' he will never be promoted to the position
of president himself, and feels 'a little blue about my future'. He
closes by saying that Ismay's 'position regarding the *Titanic* is
improving every day', and that he has 'absolutely nothing to
reproach [himself] with. You were saved for some good purpose
and must take advantage of it.'[7]

In August, Ismay writes again to Marian Thayer. Still unable to
face work, he has been shooting in Yorkshire for three weeks and
is about to return there. 'I love it so. The heather and grass.' But
despite the isolation, he has not been happy. 'I always have the
memory of that awful time before and cannot think of anything
properly. How I wish the hands of the clock could be put back.
Think of all our ambitions ended – at least mine are – and all one's
work ruined.'

Ismay tells everyone, not only Mrs Thayer, that the first half of his
life is now meaningless. William Boulton, a soldier who has known
him for years, tells Ismay to 'not say again your life is ruined – don't
say it, or even admit it to yourself. You are still suffering from nerve
shock which would have broken most men down and you are not
competent to judge.' It is vital, Boulton continues, to stop all this
'morbid introspection'; to let 'your friends estimate your worth and
tell you what value they place on your life's work'.

Ismay receives other letters of consolation: Mary Anna Maxwell
Stuart, the sister of an old friend and the great-great-granddaughter
of Sir Walter Scott, tells Ismay 'how perfectly frantic with indignation we are at the way you have been treated! As if anyone who had
ever known you could have believed you capable of any mean
action!'[8] Captain Mark Kerr writes on Admiralty notepaper to
express his 'indignation at the unjust, abominable and lying accu-
sations against you in some of the press and by some ignorant

people. Anyone who knows you and knows the sea, and ships, and occasions of excitement there, will see that your behaviour was most proper in every way.' He continues on a more contemporary note: 'Hysteria is the prevalent disease now – everything is at extremes . . . The wireless operator who does what he is paid for, is a hero if he does it and a miserable skunk if he doesn't! Those are the only two words they know! It makes a lot of us sick to see the stuff written.' The Commercial Club in the town of Ismay, Montana lets him know that they have no intention of changing their name. William Molson MacPherson, who formed the Canadian Dominion Steamship Company, reminds Bruce of the first disaster Thomas Ismay endured in 1873, when the *Atlantic* hit a rock and 250 people were drowned: 'History has repeated itself for your poor father went through the same distress; it prostrated him for years, but he lived and continued his wonderful work, which you are spared to continue.' Mary Jones, who remembers Ismay as a child in Liverpool, writes that as a nurse during the Boer War she gained 'a very intimate knowledge of men when in extremity. I discovered that the greatest heroism and bravery are not by any means always associated with the truest chivalry – often far otherwise. Personally I think to live is often much harder than to die – I do not doubt you will agree . . . you have many unknown friends, believe me.' Ismay's sister ticks him off for stopping his Sanatogen – 'You could not begin to feel the effects for at least EIGHT WEEKS. I know it will do you good – try it again.' Harold Sanderson begs him to 'try to see things as they really are, and not through glasses of a morbid tint . . . I am confident that you are wrong to worry about your own position so much. This is', Sanderson concludes in a Jamesian vein, 'a very interesting experience.'[9]

Two people who Ismay hears nothing from are J. P. Morgan and William Pirrie. 'As usual,' Pirrie's wife explained later in the year, 'when feeling strong and sorry over a matter [William] is very quiet and sad over it.'[10]

In her last letter, Marian Thayer says that she is getting out more and visiting friends, that she is constantly surprised by how good and

kind everyone is to her. Why be surprised? Ismay replies. 'It is the
most natural thing in the world.' She wonders if he believes the verse
she sent him, which he placed on his dressing table? He says that he
does, and that 'doing so is a comfort'. She offers to send him a book
which is filled with similar verses and prayers; he says he would appre-
ciate it. She asks him, as she also asked President Taft, whether he
believes in life after death? Ismay replies that he has 'several times
heard my mother's voice but I think it must only be in my dreams.
How I loved her.' He does not, he admits, 'believe in reincarnation
but I feel as you do in regard to knowing each other'. They had, Mrs
Thayer suggested, met in a previous life. 'There are a great many
things I should love to talk to you about and some day,' Ismay continues, 'we will have the opportunity of doing so.' He then repeats that
'I cannot blame myself in any way for the awful disaster', adding that
he 'had no more to do with it than you had'. It's a startling remark by
any measure, and it must have startled Mrs Thayer. That Ismay is as
blameless as she is, or she as culpable as he is, had not occurred to
her. As far as Marian Thayer is concerned, while Ismay should not be
expected to shoulder the entire responsibility for the disaster, he must
surely – as a man of honour, a businessman of repute, and a church-
going Christian – acknowledge some responsibility. To err is only
human. Had he learned nothing from the two inquiries?

Her sympathy has already started to wane, and she forgets to
enclose her photograph in the letter. She is filing an insurance claim
for the loss of her husband's life and asks Ismay whether he might, as
a favour, present her case to Mr Franklin for special consideration.
Ismay replies that while 'I would do a great deal for you, and you
know it, what you ask is impossible.' He is 'satisfied you really don't
mean what you write. I am deeply sorry for the loss you have
sustained and of course I know any claim you put in would be absolutely right, but you must agree with me that all claims must be dealt
with on the same basis now don't you?' Mrs Thayer is doubtless used
to being patronised, but the sudden imposition of boundaries and
proprieties in a relationship which has so far dispensed with them
altogether comes as a further surprise. Ismay begins his letter by

suggesting that he and she were equal in their blamelessness; he agrees that they knew one another in a previous life; he is resting his future on her goodness; he guesses that she alone stood up for him on the *Carpathia*, that she prevented Mrs Ryerson from mentioning him in her affidavit, that she understands his pain, but he will not extend his influence as her friend. 'You must not think me unkind or inconsiderate,' he continues, 'and I am satisfied that if you will think this matter over you will agree that I am acting rightly. Don't let us say anything more about this please.' No more was said and Marian Thayer, who asked the White Star Line to compensate her for the loss of her luggage, never claimed for the loss of her husband.

Ismay then returns to the perennial subject of whether the two of them will ever meet again. 'Of course we will,' he says, 'why do you get such ideas in your head? Writing is a very poor substitute but surely it is better than nothing so don't make this an excuse for not writing. You will surely be coming to this side ere long and we will meet.' He knows he will never again go to America. He ends in his usual fashion, apologising for 'such a long letter' and telling her she is 'constantly in [his] thoughts'. In a postscript he adds that 'it is always difficult to say what one would like at the end of a letter but I'm sure you will understand'. Mrs Thayer understands everything.

To and fro the letters go, each one trying to replicate the intensity of their first contact. Ismay repeatedly describes himself as 'talking' to Marian Thayer but it is remarkable how little he says beyond expressing his desire to talk to her further. What, apart from his sense of blamelessness, does he want to talk about? Does he want this to be an amorous exchange, or one in which they simply comfort one another as fellow beings? Are his reasons for writing the same as her own? Whatever it is he is trying to say, his words simply freeze in the air before melting. He expresses his trauma as a confined banality and his letters, like his responses to the two inquiries, wander around and around the same immovable fact. He plays along with Marian Thayer's peculiar fads in which he clearly has no real interest, he takes on board her specific vocabulary as though he

has no words of his own, and he hands back nothing but the continued drip of his sense of unfairness. When Ismay next writes it is
mid-December, he has turned fifty and the shooting season, which
has taken 'my thoughts from other things', is nearly over. Apart
from his daughter Margaret, 'who has left her husband to keep me
company', he is alone in the house. Florence and the children are
currently in London, but will be returning to Liverpool for
Christmas when they are expecting a 'large party – all family, sisters,
brothers, son-in-law'. Ismay is dreading it. 'I am not fond of the
so-called festive season, full of pleasure for the young and full of
sorrow and unhappy memories for the old like myself.' Forgetting
that he has already, in a previous letter, confided in her about his
retirement plans, he now confides again, 'in the strictest confidence',
that he is 'going out of business on 30 June' and that this decision
has nothing to do with 'that terrible calamity'. He will be sorry to
no longer be involved with the White Star Line as he 'loves the business and all connected with it' but he has 'neglected other matters'.
The reason for his early retirement, he reveals, is to 'see more of my
family and to try and make them happy', adding that 'it will be a
very great wrench. I will leave Liverpool and settle somewhere in the
country.' But can this really be the reason? Ismay's children, aged
twenty-three, seventeen, fifteen and ten, are now drifting in their
own directions and his marriage has been over for years. Ismay,
whose motivations always have a secret side, has been divesting
himself of his chattels since the day his father died.

He is more anxious about the future than he is about the past; as
usual he is looking straight ahead. He is moving into a life of being
rather than doing and his anxieties about retirement are nothing
new; before he boarded the *Titanic* he had expressed to Harold
Sanderson his 'very mixed and doubtful feelings' about the coming
years. He delayed the date of his retirement from the presidency of
the IMM and the chairmanship of the White Star Line from January
to June 1913 because he thought it would be easier in the summer
months to deal with the 'little or nothing' he now had 'to fill up my
time'. He now has cold feet again about giving up the White Star

Line, a company for whom he feels a 'sentimental attachment', and announces that he has changed his mind about retirement, that while he is willing to let the IMM presidency go to Sanderson, he still wants to be at the helm of his father's firm. J. P. Morgan replies this will not be possible; that Ismay has, 'by your ability and your strong personality overshadowed the other managers' and that it would be 'easier for these men, as also for the incoming president, to assert their independence if their former chief is not on the boards with them'. Ismay does not understand what Morgan is saying to him; that he is considered a liability, that he represents the errors of the past, that he is seen as an autocrat around whom no one can breathe.[11] He does not mention in his letters to Mrs Thayer that he has been excised from his own company.

Marian Thayer's most recent letter to Ismay had contained a selection of photographs of herself and Jack, and Ismay has chosen to keep two of the mother and one of the son – 'I hope you will not consider that I am too grasping.' 'Do you know,' he continues, looking at her picture as he writes, 'not a single day goes by without my thinking of you. I cannot understand it.' He wonders how she is and wishes he could talk to her: 'How I should love it.' He then, as though it were the most natural thing in the world, says something extraordinary: 'I often think of where our friendship would have taken us if that awful disaster had not taken place, how well I remember our conversations when everything looked so bright. You had a very peculiar attraction to me and I loved talking to you and hearing you talk, you interested me so much . . . How I ramble on. I must stop.' Had the *Titanic* not gone and hit the iceberg, he and Mrs Thayer could have been happy together: this is what occupies Ismay's mind. 'My God,' as Jim says to Marlow when he jumps from the *Patna*, 'what a chance missed, what a chance missed.'

In a newspaper article written on the twentieth anniversary of the wreck, Jack Thayer recalled that when he was standing on the deck with his parents, Ismay approached them and told them that the ship had one hour to live. Ismay, who did not even warn his own valet, of course wanted Marian Thayer to live and of course wanted to live himself, especially now that there was someone worth living for.

———————

In October, six months after the disaster, a collection of three sea stories by Joseph Conrad appeared in the shops. Titled *Twixt Land and Sea,* his new book marked a turning point in Conrad's career. No longer under the shadow of Henry James, Conrad was once more the master of 'adventure and psychology', he was flying his 'old colours of mystery, romance and the strangeness of life'.[12] It was the second story in the volume, called 'The Secret Sharer', but variously titled in draft 'The Secret Self', 'The Second Self', 'The Other Self' and 'The Secret-Sharer', which was his favourite. It appeared, perfect and complete, in two weeks; Conrad experienced none of the usual agonies and interruptions in the writing process. Some part of him had been released by the tale; it made him feel curiously elated. '"The Secret Sharer", between you and me, is *it.* Eh?' he wrote to Edward Garnett. 'No damn tricks with girls, there, eh? Every word fits and there's not a single uncertain note. Luck, my boy. Pure luck!'

'The Secret Sharer' is told by a fledgling captain who remains unnamed. He has been in command of his ship – also unnamed – for under two weeks and is still waiting to see whether he 'should turn out faithful to that ideal conception of one's own personality every man sets up for himself secretly'. He is soon to find out: keeping watch in his sleep suit one night, rejoicing 'in the great security of the sea as compared with the unrest of the land', he finds in 'the sleeping water' a naked man hanging onto the rope ladder.

The Captain pulls the man, whose name is Leggatt, on board. Like Jim, Leggatt is the son of an English country parson. He has swum several miles from his own ship, the *Sephora,* where, during a storm, he landed a blow on the head of a mutinous seaman. The seaman is killed, and Leggatt kept prisoner in his cabin. After nine weeks of incarceration, he is overcome by 'a sudden temptation' and jumps: 'I was in the water before I had made up my mind fairly', and he swims until he sees the lights of this present ship, when he decides to chance his luck. Leggatt feels blameless: in killing the man he saved his ship, and he has nothing but disdain for the law by which he will be judged should he ever again see land. 'But you don't see me coming back to explain such things to an old fellow in a wig and twelve respectable tradesmen, do you? What can they know whether I am guilty or not – or of *what* I am guilty, either?' The Captain feels sympathy for the fugitive, who is older than him by five years and trained, as he did (and as Captain Rostron did too), on the HMS *Conway* in Liverpool (a training ship which was also a concern of Thomas and Bruce Ismay). He gives Leggatt a sleeping suit of the 'same grey-striped pattern as the one I was wearing' and, at the risk of losing his own authority with the crew, keeps him hidden in his cabin. 'He appealed to me as if our experiences had been as identical as our clothes . . . I saw it all going on as though I were myself inside that other sleeping suit . . .' Leggatt is the Captain's 'double', his 'reflection in a dark and sombre mirror'. Conrad forces on the reader the fact of the Captain's identification, using the word 'double' on nineteen occasions. 'I knew well enough . . . that my double there was no homicidal ruffian. I did not think of asking him for details, and he told me the story roughly in brusque, disconnected sentences. I needed no more.' If anyone were to see the pair of them, 'he would think he was seeing double, or imagine himself come upon a scene of weird witchcraft; the strange captain having a quiet confabulation by the wheel with his own grey ghost'. Not that the two men looked in any way similar. 'He was not a bit like me, really; yet, as we stood leaning over my bed place, whispering side by side, with our dark heads together and our backs to the door, anybody bold enough to open it stealthily would have been treated to the uncanny sight of a double captain busy talking in whispers with his other self.' Leggatt too has found his sharer: 'As long as I know that you under-stand,' he whispers. 'But of course you do. It's a great satisfaction to have got somebody to understand. You seem to have been there on purpose . . . It's very wonderful.'

The Captain's description of himself with his looking-glass reflection recalls the picture Conrad painted of himself and Marlow, 'when, in silence, we lay our heads together in great comfort and harmony'. But the company of Leggatt does not offer comfort and harmony: the Captain is 'creeping quietly as near insanity as any man who has not actually gone over the border'. In the story's thrilling final scene the Captain takes a risk with his ship and 'shaves the land as close as possible' to allow Leggatt to leg it and swim to the island of Koh-ring. 'The secret sharer of my cabin and of my thoughts, as though he were my second self, had lowered himself into the water to take his punishment: a free man, a proud swimmer striking out for his new destiny.'

'The Secret Sharer' belongs to the same world as *Lord Jim*: there is a crime at sea, a jump, a story to unburden, an understanding captain, and a second chance. Both Jim and Leggatt commit a 'breach of faith with the community of mankind', and both men make sense to themselves only in relation to another. 'We exist,' as Conrad puts in, 'only in so far as we hang together.' And like *Lord Jim*, 'The Secret Sharer' is based on a real event. In September 1880, during an argument on board the *Cutty Sark*, the chief mate, Sydney Smith, brought a capstan bar down on the head of a black seaman, who died from injuries three days later. Sydney Smith persuaded the Captain to let him jump overboard where he swam until he was smuggled onto another ship. In his shame, the Captain of the *Cutty Sark* then killed himself. In Conrad's hands, the scandal, which was reported in every paper and discussed in every port, becomes the tale of two men in their sleep suits whispering in a ship's cabin, and this dreamlike scene recalls Ismay in his pyjamas, hiding in the doctor's cabin on the *Carpathia*, plotting with Lightoller how to escape the inquiry without anyone noticing. Lightoller, like Conrad's captain, is caught between personal conscience and corporate duty. Will the officer prove 'faithful to that ideal conception of one's own personality every man sets up for himself secretly'?

Lightoller (who shares his own secrets with his wife, whom he met on board a White Star ship) is not Ismay's only secret sharer. Wherever we find him, Ismay is one half of a whispering partnership. He whispers side by side with Harold Sanderson, his secret replacement at the IMM; he whispers over dinner with Lord Pirrie when they plan to upstage the *Lusitania* with the *Titanic*; he whispers with William E. Carter, who jumped into his lifeboat at the same time; with Captain Smith, who handed him the Marconigram from the *Baltic*; with J. P. Morgan, whose involvement in the finances of the White Star Line would come as a shock to the British inquiry; and most importantly he whispers with Mrs Thayer. He is mired, Ismay reveals to her, in the moment of his jump; he is lost and can only find himself again in her. Ismay and Marian Thayer form, he feels, a part of one another. It seems that everyone, except Florence, is Ismay's secret sharer.[13]

To Ismay's family it was Kipling and not Conrad who best described his internal struggle, and Bruce's sister-in-law sent him a copy of Kipling's 'If' because 'it reminds me so of you'. First published two years earlier in the collection *Rewards and Fairies*, Kipling had written the verse in 1895 in celebration of Dr Leander Starr Jameson, a British colonial statesman who, with the secret support of the British government, had assembled a private army outside the Transvaal to overthrow Paul Kruger's Boer government. The uprising, known as the Jameson Raid, was prevented by the Boer forces and Jameson was forced to surrender. His much-publicised trial took place in London where he was lionised by the press as 'the hero of one of the most daring raids in all the annals of border warfare'. Jameson held his tongue throughout about the involvement of the government under whose auspices he was now being tried, and presented himself in court as 'a quiet, modest gentleman, in faultless and fashionable dress, with civilian stamped upon him from head to foot'.

> *If you can keep your head when all about you*
> *Are losing theirs and blaming it on you,*
> *If you can trust yourself when all men doubt you,*
> *But make allowance for their doubting too;*
> *If you can wait and not be tired by waiting,*

Or being lied about, don't deal in lies,
Or being hated, don't give way to hating,
And yet don't look too good, nor talk too wise:
If you can dream – and not make dreams your master;
If you can think – and not make thoughts your aim;
If you can meet with Triumph and Disaster
And treat those two impostors just the same;
If you can bear to hear the truth you've spoken
Twisted by knaves to make a trap for fools,
Or watch the things you gave your life to, broken,
And stoop and build 'em up with worn-out tools –

If you can do these things and more, Kipling concludes, 'you'll be a Man, my son'. To his family, Ismay was shouldering the blame with the stoicism of a soldier.

———

Ismay's next letter to Marian Thayer is written on the last day of the year. Their previous letters have crossed in the post; she has sent him Emerson's essay on 'Compensation' and two books of prayer, *Out from the Heart* and *I Thank Thee*. They all, Ismay says, 'contain so much that is true and helpful'. She has also written to Florence to 'make inquiries' about Ismay's health following a story in the American press. The article was groundless – based, Ismay supposes, on rumours about his imminent retirement – but Mrs Thayer, who now realises that Ismay has misunderstood their relationship, is in rapid retreat. He must not be allowed to talk about his 'attraction' to her or 'where our friendship would have taken us' had the ship not hit the iceberg. As far as she is concerned, they are simply two unhappy survivors who have shared a ghastly experience. Florence is pleased to hear from Mrs Thayer and shows her husband the letter. It is unlikely that Ismay reciprocates the gesture.

Christmas with the Ismays is particularly grim. He finds it 'impossible to get through' and imagines that Marian will be feeling the same. She was in his thoughts 'constantly . . . I wonder if you felt it?' He is filled with self-loathing, and now that Mrs Thayer has woven his wife into her world of connections he feels licensed to talk about the difficulties of his home life. Rather than creating a distance between herself and Ismay by writing to Florence, Mrs Thayer has inadvertently encouraged Ismay to clear his throat and move the conversation on. He has, he now reveals, 'such a horrible, undemonstrative nature. I cannot show people how fond I am of them and not doing so hurts their feelings and they imagine I do not care for them. What can I do? Of course, one cannot hurt other people without hurting oneself.' It is unlikely that the account of his failings he gives Marian Thayer has ever been expressed to Florence, that Ismay has ever told his wife how agonising he finds it to watch himself hurt others or that the silent person she is currently living with bears no relation to the person he feels himself to be. 'Very often,' he continues to Mrs Thayer, 'a word would make things right, one's horrid pride steps in and this causes unhappiness. I wonder if you know all I mean.' Mrs Thayer, who seems never to have suffered from horrid pride, may not know all he means, as Ismay suddenly sees. 'I can hear you saying what a horrible character and I agree. I absolutely hate myself at times. Tell me what I can do to cure myself.' She has told him in an earlier letter not to 'lock' himself 'up tight'. Ismay is amazed that Mrs Thayer can have picked up on this aspect of his character: 'When did you notice that this was another of my failings? Do you know that I always put my worst side forward and very very few people ever get under the surface? I cannot help it, can you help me to change my horrid nature?' He still has 'awful fits of depression' and wishes he 'could make something of the life that remains'. Perhaps, he now writes to Mrs Thayer, 'something may turn up. My wife says it will but I am very disheartened.' He asks to be remembered to Jack, and wishes that he could 'have a talk with you. It would be a great help to me.' While Ismay has not yet been over to Ireland to see his new house, a friend visiting on his behalf reports that the place is primitive and unmodernised and needs a good deal of work.

In February the nation is gripped by the tale of another voyage out when the body of Captain Scott is found frozen in the Antarctic. The explorer was, according to the *Daily Mail*, 'a true and spotless knight' and the journey to the South Pole an example of 'modern chivalry'. The failure of Scott's party to be first to reach their goal is redeemed by the nobility of their deaths. Ismay must seem the only Englishman to ever return from a voyage alive. He continues to send Mrs Thayer letters on the White Star mail ships, but she now no longer replies. By the end of the month Ismay tells her that 'it is ages and ages since I have had a word from you'. He has no idea of her 'doings' or her plans for the summer. Perhaps, from what Florence tells him, she has 'not been very well lately?' Marian corresponds with Florence more than she does with Bruce. He tells her that he has been over to Ireland to see the house, which he fears that Florence, who loves London, will find dull. Never mind. He can personally no longer bear the city, and they have let their London house until the end of July. He is currently in Liverpool, where his 'business career is drawing to its close'. He still has 'mixed feelings' about life without work.

He next hears from Mrs Thayer in April, a week after the first anniversary of the *Titanic*'s sinking. He has by now 'almost given up waiting'. She has been unwell and is thinking of going to Switzerland; she asks if he has read the books she sent him and which ones he likes best? He tells her that 'they were all so nice and comforting. You have been much in my thoughts.' He asks God in his prayers to give her strength to bear her loss. 'It is awfully difficult at times to realise the awful thing that has happened.' He hopes she might pay him a visit if she is in Europe, as 'I would dearly love to have a good talk with you.' He has not been doing much; one of his nieces is getting married. 'I have to go but there is nothing I dislike more than a wedding.' He is looking ahead to life in Galway and wondering how to 'occupy my time when I go out of business'. It still 'makes me very unhappy to think that I am severing my connections with a concern made by my father the most successful in the world, one of which I was so proud. I loved the ships.' He fears he is 'boring' her and says she is 'always in my thoughts'.

She replies almost instantly with words of wisdom and sympathy, and he reads her letter many times: 'it has helped me so much'. His own reply is filled with self-recrimination: how do his friends put up with him? He is so anti-social, he doesn't like people or crowds. He has been 'shooting' and 'praying', she is 'much in my mind', she knows his faults better than anyone else, only she understands that he is someone who 'feels very deeply and [is] extremely sensitive and undemonstrative. I cannot show myself as I really am and always put my worst side forward. I don't know why I unburden myself to you in this manner.' Mrs Thayer waits for a month before replying with a cable, asking him to remind her of the names of the three books she had sent him, which she is evidently now recommending to someone else. 'Please do not think a cable message is satisfying,' Ismay responds, 'you owe me at least two letters and they can't be cables.' He is going to Ireland this week and 'looking forward to getting away and living in the country. I'm sorry to say I never liked people and am now worse than ever in this respect.' He hopes that Florence will not be 'bored to death', otherwise he will have to sell the house. He has been fishing in Dorset, playing a good deal of golf ('I know of no game that takes one so out of oneself') and in August will go shooting in Scotland. His philandering brother, Bower, won the Derby, to then have the prize 'taken from him owing to the wrong riding of the jockey. Everybody feels he has been most unfairly treated.' Will Mrs Thayer ever be 'coming to this side?'

Here the correspondence, as it exists, comes to an end. Perhaps they continued to write and the letters are lost, but it seems unlikely. They had nothing further to say to one another and Ismay, who told his story to Mrs Thayer as best he could, gave up the hope that he would one day be able to talk to her in person. Instead, he may have had an affair with another *Titanic* survivor, the fashion journalist Edith Russell, whose life he had saved by insisting that she get into a lifeboat when she was standing on the deck. Miss Russell, who became famous for leaving the ship with her 'lucky pig', later confided to William Macquitty, who produced the film of *A Night to Remember*, that she and Ismay were to become 'more than just friends'. There is no further evidence of their relationship, but it is easy to believe. Ismay needed a secret sharer and Edith Russell was one of the few survivors who saw him as a hero. What remains of their correspondence, however, shows that he had no particular feelings for Miss Russell in the years immediately after the *Titanic*.

Mrs Thayer came to England on several occasions but never saw Ismay again. She spent the rest of her life on the 'other side', making contact with her lost husband through her new discovery of mirror-writing.

The Super Captain

The isle is full of noises

The Tempest, III, ii

On 2 May 1912, as the *Adriatic* steamed out of New York carrying Ismay in one of its cabins, the British Board of Trade inquiry into the wreck of the *Titanic* began in London. The Scottish Drill Hall at Buckingham Gate, armoury of the Scottish Rifles, had been chosen as a venue because it could hold several hundred spectators, most of whom would be, it was anticipated, ladies with time on their hands. Described by the *Daily Mirror* as resembling 'a gigantic swimming pool', the grim building proved too capacious by far; an audience of less than one hundred, including only ten women – of whom one was Virginia Woolf – settled into the galleries for the opening proceedings. Two hundred further seats were taken on the floor by presiding lawyers and pressmen.

Now that the mariners were dry and the narrative was no longer unravelling in real time, interest in the *Titanic* was beginning to wane. The public had feasted on the tragedy for long enough; the weather was getting warmer and the forthcoming summer promised a brilliant social season. Londoners were looking for distractions, and J. M. Barrie's gift of a statue of Peter Pan to Kensington Gardens received more press attention than the inquiry's opening session. The essentials of the wreck were now widely known and the dramatis personae – Ismay, Lightoller, Rostron and Captain Smith – were familiar figures in every household. The US inquiry had been reported in full by the British press; Ismay's statement of self-defence had been read over breakfast by the whole country and his second examination by Senator Smith had been transcribed in the pages of *The Times* only the day before. The British inquiry, it was assumed, would be no more than an echo. The assessors would take the script produced by William Alden Smith and dignify it a little, introducing a new interpretation here, a different character-reading there. As it was, the British inquiry added to the mystery of Ismay's actions one vital scene which had been overlooked by Senator Smith, while they moved another from the margins of the drama to the centre.

Despite the arrival of the assessors in mourning, the inquiry began on a note of triumph rather than tragedy. Sir Robert Finlay, representing the White Star Line, announced in his opening statement that 'this disaster has given an opportunity for a display of discipline and of heroism which is worthy of all the best traditions of the marine of this country'. The following day, lookout Archie Jewell and Able Seaman Joseph Scarrott gave the first witness accounts on home ground of the wreck of the *Titanic* to a gallery now containing only one woman and a policeman. Reports of the crew's statements, described by *The Times* as 'thrilling narratives of the last scenes on the doomed ship', could be found nestling in the heart of the *Daily Mirror* while the paper's headlines announced 'Britain's First Aeroplane Warship To Take Part in Next Week's Naval Inspection By the King'.

What had drawn the crowds to the US Senate inquiry was the promise of language unrehearsed and under pressure, of voices betraying their speakers. The US inquiry had proved a conundrum; as the events of the night of 14 April unravelled, the truth became more and more elusive. Nothing made any sense: while everyone told the same story ('I went to bed, I heard a scraping sound, I went on deck, I got in a lifeboat, I was cold'), no two accounts appeared to agree. Speech became fraught with danger as witnesses tried to steer their answers around the traps placed in their way by William Alden Smith. Sentences subsided, imploded, broke down into babble. Those who had attended the British inquiry in the hope of hearing language at breaking point found that the acoustics of the Scottish Drill Hall were such that they could hear nothing at all. The proceedings, so far from the galleries that they had to be watched through binoculars, were closer to a silent newsreel or a series of lantern slides than a live event. Even the assessors cupped their hands to their ears in order to catch what was said.

The hall, which had been turned into a makeshift court with a great swath of velvet curtain covering up the brickwork, was focused around a witness stand, a twenty-foot model of the *Titanic* carrying sixteen miniature lifeboats, and a large chart of the North Atlantic. To the right of the witness stand was a dais on which was seated the Wreck Commissioner, Lord Mersey and, at a lower level, his five assessors. Facing Lord Mersey were the other members of the Board of Inquiry, who together made up the sharpest legal minds in the country: the Solicitor-General, Sir Rufus Isaacs QC; Sir Robert Finlay, Liberal Unionist MP for St Andrews and Edinburgh Universities, representing the White Star Line; Thomas Scanlan, MP for Sligo, representing the National Sailors' and Firemen's Union; and Clement Edwards, MP for the Welsh mining seat of East Glamorganshire, and representing the Dockers' Union. A sounding board, resembling an enormous box lying on its side with the lid open, was installed on the dais, but the contraption made little difference. 'Lord Mersey's questions to counsel and witnesses', reported the *Daily Mirror*, still 'sounded like whispers.' The hushed proceedings seemed symbolic of the fact, now increasingly apparent, that the witnesses had very little to reveal about what had happened to the *Titanic*. As the *Mirror* put it, 'in the midst of expectancy, as terrible almost as the silence of that frosty sparkling night on which more than fifteen hundred people went slowly into the depths, there is

absolutely no answer to give! There is no answer. There is a pitiful
stumbling of words.'

———

Lord Mersey was not the effete aristocrat the Americans, misled by
his title, assumed him to be. The son of a Liverpool merchant, John
Charles Bigham, now aged seventy-one, had grown up around
wealthy shipowners. His own father's childhood had been, as he put
it, 'grindingly poor' and before making his fortune in insurance,
Bigham Senior had worked as a clerk in a shipping office. Mersey
knew exactly where Bruce Ismay had come from; he had mingled
socially with the Ismay family; he dined at the same clubs, he
belonged in the same social rank, and in 1905 he had even spoken for
Margaret Ismay in a suit she brought against a driver who had crashed
into her Panhard-Levassor during a driving holiday in Scotland. Like
Thomas and Bruce Ismay, Lord Mersey was a Liberal Unionist who
was against Irish Home Rule, and for a brief moment he stepped
down from the Bar to represent the Exchange Division of Liverpool
in the House of Commons. In his legal career, Mersey had excelled
in representing shipping lines and the title he took on his elevation
to the peerage was no doubt the one that Thomas Ismay would have
chosen for himself had he been similarly honoured. The Mersey was
the road to Empire, Liverpool's golden seaway. He would, Lord
Mersey joked, leave the title of the Atlantic itself to one of his
esteemed colleagues.

Because the British Board of Trade inquiry into the *Titanic* was
inquiring into its own failings, it was assumed by the public that the
procedure would be self-serving and that Lord Mersey had been
chosen as Wreck Commissioner because he was an old hand at
dodging the issue. Mersey had, reported the *New York Times,* 'a
record for official blindness to well-known facts'. During his time as
an MP, Mersey had cross-examined in the House of Commons
inquiry into the role played by the British government in the South
African Jameson Raid. When the government was duly cleared,

Mersey was so generally regarded as having orchestrated a whitewash that his name became linked with the term. G. K. Chesterton, in an article dramatising the difference between the US and British inquiries, has the Englishman tell the American: 'I know you and your popular persecutions. You will hunt poor Ismay from court to court, as if he were the only man that was saved – just as you hunted poor Gorki from hotel to hotel, as if he were the only man not living with his wife.' The American replies: 'I know you and your gentlemanly privacies and hypocrisies. You will shirk this inquiry just as you shirked the inquiry into the Jameson Raid.'[1]

Lightoller later wrote in his memoirs: 'The Board of Trade had passed that ship as in all respects fit for sea, in every sense of the word, with sufficient margin for everyone on board. Now the Board of Trade was holding an inquiry into the loss of that ship – hence the whitewash brush.'[2] It was not only the public who had little confidence in Mersey as an impartial assessor. The Bradford and District Trades and Labour Council sent the Home Office a resolution of 'no confidence in the Court of Inquiry'. It was, they concluded, 'an attempt . . . to whitewash those who are the most responsible for the terrible loss of human life'. [3] When it was over, the British Seafarers' Union announced that 'the Whitewashing "Titanic" inquiry has cost the nation £20,231 . . . It will be seen that the lawyers take between them just on £13,000 . . . The ruling classes rob and plunder the people all the time, and the Inquiry has shown that they have no scruples in taking advantage of death and disaster.' As far as Conrad was concerned, Lord Mersey could be taken no more seriously than the Mikado's ridiculous Grand Pooh-Bah, who was First Lord of the Treasury, Lord Chief Justice, Commander-in-Chief, Lord High Admiral, Archbishop of Titipu, Lord Mayor, and Lord High Everything Else.

––––––––––

The response of the British press to the wreck of the *Titanic* had been to celebrate Britishness rather than allocate blame. Several papers

commented on the analogy between the heroism and discipline of
the men on the *Titanic* and those on the *Birkenhead*, and on the
alleged fact that musicians were performing as each ship went down.
The two journalists who produced the first book-length accounts of
the wreck agreed that this was to be a story without a villain, a narrative decision which left the behaviour of Ismay problematic. Filson
Young's *Titanic*, a work of high Romanticism and the first recorded
instant book, appeared in the shops on 22 May. Young declared that
'there is nothing so entirely dignified, as to be silent and quiet in the
face of an approaching horror', adding of Ismay, with Marlow-like
empathy, that he carried his cross with the stoicism of an Englishman:
'he bore on his shoulders the burden of every sufferer's grief and loss,
and he bore it, not with shame, for he had no cause for shame, but
with reticence of words and activity . . . and with a dignity which
was proof against even the bitter injustice of which he was the victim
in the days that followed'.[4] In 'The Deathless Story of the Titanic',
published by *Lloyd's Weekly News* two weeks after the wreck, Philip
Gibbs came up with the phrase: 'Yet greater than tragedy is the glory.'
The British felt more anger towards Senator Smith than they did
towards Bruce Ismay, but the general mood was one of sorrow.
'Slowly, with infinite reproach,' wrote the *Daily Mirror*, 'the whole
world turns towards those responsible and asks them, *why*? There is
no tone of vulgar recrimination, no calling of names and bringing up
of useless bitterness in this gesture. It is simply the sorrowful turning
of all those who sympathise towards those who might know, followed
by their breathless question: *Why?*'

The British inquiry divided this overwhelming question into
twenty-six specific areas, to which 25,622 answers were given by
ninety-four different witnesses. They covered the issue of safety
provision, the arrangements for the manning and launching of boats,
the route taken by the ship, her speed at the time of the collision, the
cause of her foundering, and the number and class of her survivors.
The inquiry would focus its energies on the role of the Marconigram
given to Ismay by Captain Smith, but nothing was asked about the
origin of the messages which had reassured the world that the *Titanic*

and her passengers were safe, a mystery which was currently occupy-
ing Senator Smith in New York, or the 'Yamsi' messages which
suggested that Ismay was trying to escape the US inquiry. None of
Lord Mersey's questions directly addressed the issue which would
became the dominant concern of the inquiry, the thread on which
the entire proceedings were strung: what was Ismay's status on board
the ship?

Senator Smith's report into the findings of the US Senate
Committee, in which he concluded that the presence of Ismay had
'unconsciously stimulated' the Captain to increase his speed, was
published on 28 May. 'We shall leave it to the honest judgement of
England,' Smith concluded, 'in its painstaking chastisement of the
British Board of Trade, to whose laxity of regulation and hasty
inspection the world is largely indebted for this awful fatality.'
Meanwhile, the only interruption to the greyness of the British Board
of Trade's proceedings had been the appearance, on 20 May, of the
glamorous Duff Gordons. Lord Duff Gordon had been accused of
paying the seven crew members in his otherwise empty lifeboat not
to return to pick up any of the bodies in the water. Two German
princes, the Russian Ambassador, Margot Asquith, and a group of
the *Titanic*'s stewardesses dressed in black turned up to hear him try
to clear his name.

The Duff Gordons were followed to the witness stand that day by
Lightoller, at which point the spectators drifted back out to feed the
ducks in St James's Park. Lightoller, who would answer 1,600 questions during fifty hours of interrogation, spoke with his usual fluency
and his description of the 'extraordinary combination of circumstances, which you would not meet again once in 100 years', made a
decisive impact on Lord Mersey. On the night of 14 April, Lightoller
said, 'everything was against us . . . In the first place, there was no
moon; then there was no wind, not the slightest breath of air. And
most particular of all in my estimation is the fact, a most extraordinary circumstance, that there was not any swell. Had there been the
slightest degree of swell I have no doubt that berg would have been
seen in plenty of time to clear it.' It was a magnificent description, so

magnificent that Lightoller later used it again in his memoirs. 'The disaster', he then wrote, 'was due to a combination of circumstances that had never occurred before and would never occur again.' It was language worthy of Conrad, who described the conditions of sea and sky on the night that the *Patna* had her accident as 'rare enough to resemble a special arrangement of malevolent providence'. Lightoller led his audience in the Drill Hall by the wrist into the uncanny world of the ancient mariner, who had killed the bird who brought the breeze. In Coleridge's mesmerising ballad, 'ice mast-high came floating by/ As green as Emerald', before the ship finds itself stuck on a 'silent sea', without 'breath nor motion,/ As idle as a painted ship/ Upon a painted ocean'. 'Wait a minute,' said Mersey, when Lightoller had finished his account of the climatic conditions. 'No moon, no wind, no swell? The moon we knew of, the wind we knew of, but the absence of swell we did not know of. You naturally conclude that you do not meet with a sea like it was, like a table-top or a floor, a most extraordinary circumstance, and I guarantee that 99 men out of 100 could never call to mind actual proof of there having been such an absolutely smooth sea.'

Lightoller was asked nothing about Ismay, other than whether he had seen him on board the ship. Nor was anything asked about the meetings between the two men in the doctor's cabin on the *Carpathia*. The relationship between Lightoller and Ismay, which had been of such interest to Senator Smith, was of no interest at all to the British inquiry, who were concerned only with the question of speed, visibility, and the role the Second Officer had played during the lowering of the lifeboats. Lightoller handled, he later said, 'sharp questions that needed careful answers if one was to avoid a pitfall, carefully and subtly dug, leading to a pinning down of blame onto someone's luckless shoulders'. Those shoulders, which had been luckless Yamsi's in America, were now, Lightoller believed, his. On this occasion he did not mock the court. Lightoller took his inquisition seriously. He ducked and dived and dodged the 'cleverest legal minds in England, striving tooth and nail to prove the inadequacy here, the lack there, when one had known, full well, and for many years, the ever-present

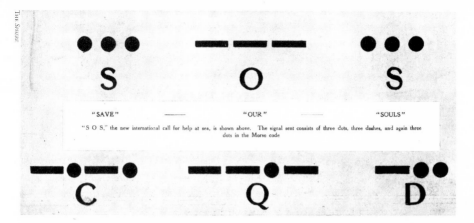

The Marconi operators, 'joked' as they sent out the distress signal CQD: 'The humour of the situation appealed to me', Harold Bride recalled, 'and I cut in with a little remark that made us all laugh, including the Captain. "Send SOS, it's the new call, and it may be your last chance to send it."'

Artist's impression drawn for the *Sphere*, April 1912. The collision occurred at 11.40 p.m.; the ship's speed was 22 knots.

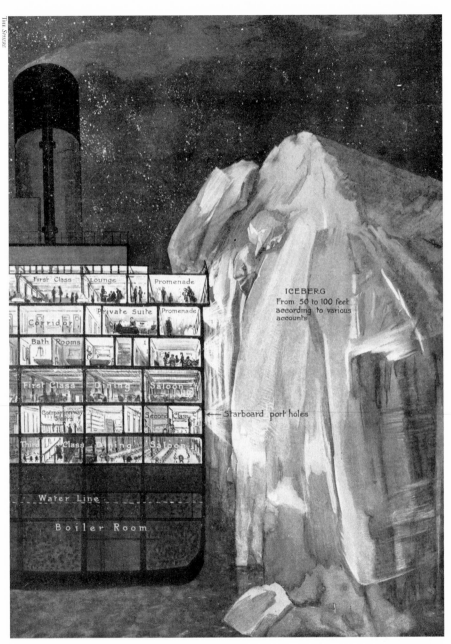

Artist's impression drawn for the *Sphere*, April 1912. The sound, one woman recalled, was like the scraping of a nail along metal; to another it felt as though the ship 'had been seized by a giant hand and shaken once, twice, then stopped dead in its course'.

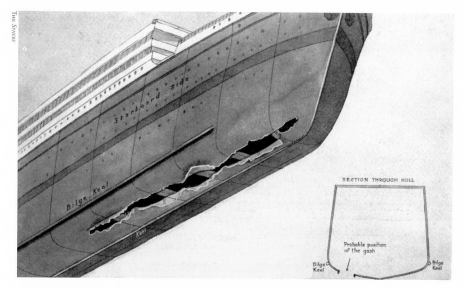

SECTION THROUGH HULL

Probable position
of the gash

Bilge
Keel

Bilge
Keel

Artist's impression drawn for the *Sphere*, April 1912. It took ten seconds for the iceberg to tear a 300-foot gash along the *Titanic*'s starboard side, slicing open four compartments.

Second Officer Lightoller on the bridge of the *Titanic*.

There were several versions of Captain Smith's final moments, in one of which he exhorts the passengers to 'Be British'.

"The Titanic looked
enormous"

Boat Deck
clear of boats

"The bows & bridge
completely under water"

Loose floating
Ice

"Sea calm as a pond
There was just a gentle heave"

Illustration by Fortunio Matania of the *Titanic*'s last moments, with annotations taken from eye-witness accounts. The *Titanic* went down two hours and forty minutes after she hit the iceberg – the same length of time as a performance in the theatre. 'I did not wish to see her go down', Ismay told the US inquiry, 'I am glad I did not.'

"A" DECK
70 feet above
the water

"B" DECK
From which many
of the women were
taken into the boats

Artist's impression drawn for the *Sphere*, April 1912. In Lawrence Beesley's account of being lowered from the *Titanic* in a lifeboat, he asks his readers to measure seventy-five feet of a tall building and then look down.

Lifeboat thought to be Collapsible C. A *Titanic* stewardess recalled Ismay 'sitting on his haunches on the stern of the boat that was cleared by the *Carpathia* ... He sat there like a statue, blue with cold, and neither said a word nor looked at us. He was nearly dead when taken on board, for he was wearing only his nightclothes and an overcoat.'

Eighteen people, including Second Officer Lightoller and Jack Thayer, survived balancing atop this upturned lifeboat.

Form No. 4.—100.—17.8.10.

The Marconi International Marine Communication Co., Ltd.,
WATERGATE HOUSE, YORK BUILDINGS, ADELPHI, LONDON, W.C.

Deld. Date. **15 APR 1912**

No. **OLYMPIC.**　　　　　OFFICE. **15 APR 1912**　19

Handed in at　**CARPATHIA**

CHARGES TO PAY.

This message has been transmitted subject to the conditions printed on the back hereof, which have been agreed to by the Sender. If the accuracy of this message be doubted, the Receiver, on paying the necessary charges, may have it repeated whenever possible, from Office to Office over the Company's system, and should any error be shown to exist, all charges for such repetition will be refunded. This Form must accompany any enquiry respecting this Telegram.

Total

To

　　COMMANDER OLYMPIC. RECEIVED 3.18 PM N.Y.T.

　　MR. BRUCE ISMAY IS UNDER AN OPIATE.

　　　　　ROSTRON.

Marconigram sent by Captain Rostron of the *Carpathia* to Captain Haddock of the *Olympic* on the morning the *Titanic* survivors were rescued.

Rostron had been Captain of the *Carpathia* for three months when he rescued the *Titanic* survivors. It had been 'absolutely providential', he said, that the modest Cunarder picked up the mighty White Star Liner's distress call.

The deck of the *Carpathia*. The *Carpathia*'s female passengers formed a relief committee to provide clothing for those *Titanic* survivors who had arrived in their dressing gowns, and ship's blankets were cut up to make warm coats for the children. One survivor later said that on the *Carpathia*, 'I learned a great deal of the fundamentals I have built a happy life on, such as faith, hope, and charity.'

When the *Carpathia* arrived at 4.30 a.m, expecting to find a damaged ship, there was nothing to see but boxes and coats and what looked like oil on the water.

On 19 April, the morning after the *Carpathia* had landed in New York, the US Senate Inquiry into the sinking of the *Titanic* opened at the luxurious Waldorf-Astoria. Ismay, seated to the right at the top of the table, surrounded, was the first witness. Headed by Senator William Alden Smith, the Inquiry produced the official version of the story of the *Titanic*, a narrative which would unfold over eighteen days, fill 1,100 pages of testimony, and destroy the reputation of J. Bruce Ismay.

DAILY SKETCH.

No. 991—MONDAY, MAY 13, 1912.　　　THE PREMIER PICTURE PAPER.　　[Registered as a Newspaper.]　ONE HALFPENNY.

MR. BRUCE ISMAY WELCOMED HOME AGAIN WITH CHEERS.

Mr. Bruce Ismay received a warm welcome from a crowd assembled on the Liverpool landing-stage on Saturday when he came ashore from the White Star liner Adriatic, in which he returned to England from New York along with the surviving officers of the Titanic. (1) Mr. and Mrs. Ismay smiling in response to sympathetic cheers as they descended the gangway. (6) Crowd cheering Mr. Ismay (in car). Mrs. Ismay met her husband at Queenstown. Below is Sir John Hare returned from Canada. (2) Mr. H. G. Lowe, the fifth officer of the Titanic, with his father and sister. Mr. Lowe was the officer who fired his revolver. (3) Mr. H. J. Pitman, third officer, with relatives. (4) Mr. C. H. Lightoller (in bowler hat), second officer. (5) Mr. and Mrs. Ismay approaching motor.—Daily Sketch Photographs.

Hounded and vilified in America, the British Press, indignant at his treatment across the pond, gave Ismay a warm welcome on his return. Florence can be seen in the two left-hand pictures; Lightoller, in a bowler hat, is in the top right photograph.

through, and how wonderful
you were through it all.
The notoriety we all got. and
the dreadful things our press
is allowed to say in this
country is certainly revolting.
and makes us sometimes
ashamed that we live here.
but fortunately, when they go
to extremes, it is quickly
over, & how it has completely
died out, & no one even
mentions it.

GWEDNA
BRYN MAWR
PENNSYLVANIA

May 24th

Dear Mr Ismay.
     I want to write you
how glad I am that you are
home safely. and also how
pleased we were to read
of the great ovation you had
in England when you landed
for no one realized more
than Billy & I did, how
much you had been

Letter to Ismay from Lucille Carter, whose husband William E. Carter also jumped into Collapsible C but whose survival was of no interest to the press or inquiries. Mrs Carter would later divorce him on the grounds that he had abandoned her on the *Titanic*.

Signed photograph of Marian Thayer, with whom Ismay had fallen in love on the *Titanic*: 'I often think of where our friendship would have taken us if that awful disaster had not taken place', he would write to her the following year.

In Fortunio Matania's illustration of the British Inquiry Ismay has massive presence and no presence at all, he is both the smallest and the largest man in the room. In contrast to the crowded table in the Waldorf-Astoria, the British Inquiry was grey and formal. The accoustics were so bad that the assessors had to cup their hands to their ears to hear what Ismay said.

Bruce and Florence
outside their house in
Costelloe, Connemara,
where Ismay was
believed to be 'living
among the missing'.

Ismay's garden
at Costelloe was
designed by Gertrude
Jekyll. 'It is awfully
wild and away
from everybody,'
Ismay wrote to Mrs
Thayer. 'I will enjoy
the place.' His last
summer in Ireland
was in 1936.

possibility of just such a disaster'. He was proud of his performance, which he called a 'long drawn out battle of wits', and he was keen afterwards to let it be known that he held 'the unenviable position of whipping boy' for the White Star Line. He defended the company from the accusation that the ship was undermanned when it came to launching the lifeboats; he stated that it was common practice not to slow down in ice regions; if it was considered reckless to head into an ice field at full speed, Lightoller said, then 'recklessness applies to practically every commander and every ship crossing the Atlantic Ocean'.

Of the ship's movements after the collision, he repeated what he had told the US inquiry: that the engines had stopped while he was lying in his bunk, he had gone to the bridge in his pyjamas and seen that the ship was slowing down, he had returned to his bed for between fifteen and thirty minutes (he told the US inquiry it was ten minutes) when he was roused by Boxhall, who told him that the water had reached F Deck. By the time Lightoller was up and dressed and on the deck, the engines had stopped completely and the boilers were letting off steam. If he was covering up the knowledge that the ship had previously been going 'Slow Ahead' under Ismay's orders, Lightoller did it without obvious effort. He afterwards said he felt more like 'a legal doormat than a mailboat officer', but the man who was appointed as the *Titanic's* First Officer to then be demoted to Second Officer, was treated by the inquiry as though he had been the Captain.[5]

———

By the time Ismay took to the stand at midday on 4 June he was no longer the most talked-of man in all the world. Talk was now of the royal visit to the zoo and the rain on Derby day, but for Ismay the hours of talk that his jump had generated hung on in the air like an aura. 'It was clear', the *Telegraph* wrote, 'that the appearance of so important a witness was wholly unexpected by the general public, because the attendance was one of the very smallest which the inquiry

has yet brought together.' A 'ripple of mild excitement' went through
the hall at the thought that the investigation might at last be 'lifted
out of the severely judicial groove' in which it had been gently
moving, and 'directed into more or less personal channels'.

This was Ismay's second chance. The US inquiry had been a dress
rehearsal; what mattered was his reputation amongst Englishmen,
particularly those in Liverpool who remembered him as his father's
son. Wearing a dark grey suit and black tie, he was, the *Daily Sketch*
noted, 'visibly nervous', but he was in marginally better shape than
he had been in America. Florence was with him; his London home
was a pleasant walk through the park from the Drill Hall, and he had
not been required to sit in the audience for the previous month as a
great narrative web spun itself around him.

Bruce (smoking) and Florence Ismay with Harold Sanderson,
walking to the British Inquiry

The American press had described Ismay as a figure of regal impe-
riousness but no one looking at him today, wrote the *Daily Sketch*,
'would have pictured this man as the head of one of the wealthiest
and most powerful shipping corporations in the world. He looks and
speaks so unlike the commonly accepted type of commercial monarch
as could well be conceived. A cultured cosmopolitan if you like, but
not a strong ruler of men.' Often standing with nothing to do as
Lord Mersey and his counsel discussed his answers amongst themselves, Ismay folded his arms, clasped his hands behind his back,
slipped his left hand in and out of his pocket, and fiddled with an
open foolscap envelope. The *Telegraph*, which thought that Ismay
showed 'no trace of nervousness', suggested that for a man who
'claimed no higher status on the *Titanic* than that of a passenger, he
looked just the part'.

Sir Rufus Isaacs began by questioning Ismay about the relationship between the White Star Line and the IMM. Although these
ships, including the *Titanic*, are registered under the British flag,
they are in fact American property? Ismay replied without hesitancy
in what was reported to be a 'low, well-modulated voice' that he had
not the slightest idea. This was an example of what the *Telegraph*
referred to as his disarming 'modesty'. Astonished to learn that the
White Star Line was no longer in any real sense a British company,
Lord Mersey then asked the witness to explain the object of an
American company managing its affairs through the English laws
affecting English companies: Why do they do it? Ismay replied that
I am afraid I cannot answer that question, My Lord. I should think,
intervened Isaacs, you ought to know. Like Lord Mersey, Rufus
Isaacs was familiar with Ismay and his background. The son of a
prosperous East London Jewish fruit merchant, Isaacs had left school
aged fourteen, became a ship's boy at sixteen, sailed to Rio and
Calcutta, briefly jumping ship mid-journey, after which he read for
the Bar and joined the House of Commons as Liberal MP for
Reading. His one year at sea left him with an abiding love of the
ocean and its traditions. A good friend of Lord Mersey's, Isaacs was
yet to be promoted to Attorney General, Lord Chief Justice,

Ambassador to Washington, Viceroy of India, and Foreign Secretary. When he died in 1935, he was Marquess of Reading, the first commoner to receive such a promotion since Wellington. Now aged fifty-two, he was enjoying one of the most spectacular legal careers of his generation. Sir Rufus Isaacs was, according to Margot Asquith, the most ambitious man she had ever known.

Isaacs then asked Ismay about his position on the *Titanic*: You sailed in her as a passenger? *I did*, Ismay replied. Did you communicate with the Captain on the fateful evening? *No*, Ismay said, *I never spoke to him at all; I had nothing to do with him at all.* You travelled as a passenger because of your interest in the vessel and in the company which owned it? *Naturally I was interested in the ship.* I mean, you had nothing to do in New York; you travelled because you wanted to make the first passage on the *Titanic*? *Partly, but I can always find something to do.* I mean to say that you were not travelling in the *Titanic* because you wanted to go to New York, but because you wanted to travel upon the maiden voyage of the *Titanic*? *Yes.* Ismay's answers were crisp and clear. Because in your capacity as managing director or as president of the American Trust you desired to see how the vessel behaved, I suppose? *Naturally.* And to see whether anything occurred in the course of the voyage which would lead to suggestions from you or from anybody? *We were building another new ship, and we naturally wanted to see how we could improve on our existing ships.* It was all, according to Ismay, in the natural course of things. Later in the day, Lord Mersey asked Ismay to 'paraphrase "naturally" and tell me exactly what you mean by it?' Ismay was unable to explain exactly what he meant, but he was using the word to mean 'logically', or what could be reasonably assumed. So, continued Isaacs, improving on future ships was the real object of your travelling on the *Titanic*? *And*, corrected Ismay, *to observe the ship.* What I want to put to you is that you were not there as an ordinary passenger? *So far as the navigation of the ship was concerned, yes . . . I looked on myself simply as an ordinary passenger.* Did you, intervened Lord Mersey, who called a spade a spade, pay your fare? *No*, admitted Ismay, *I did not.* There was general laughter. You recognise, said Sir Rufus Isaacs, that My Lord's question is one which rather disposes of the ordinary passenger theory, does it not?

The Marconigram from the *Baltic*, which reported 'a large quantity of field ice today in latitude 41.51 N., longitude 49.52 W.', was then read aloud. 'We attribute very great importance to that particular message,' said Isaacs, who would turn the Captain's handing of the Marconigram to Ismay into a cordoned-off crime scene. 'We think it is of very great importance. Now what I want to understand from you is this – that the message was handed to you by Captain Smith, you say?' *Yes*. Handed to you because you are the managing director of the company? *I do not know*, said Ismay. *It was a matter of information*. Information, suggested Isaacs, which he would not give to everybody, but which he gave to you . . . he handed it to you, and you read it, I suppose? *Yes*. Did he say anything to you about it? *Not a word*. He merely handed it to you, and you put it in your pocket after you had read it? *Yes, I glanced at it very casually. I was on deck at the time*. Had he handed any message to you before this one? *No*. So this was the first message he had handed to you on this voyage? *Yes*. And when he handed this message to you, when the Captain of the ship came to you, the managing director, and put into your hands the Marconigram, it was for you to read? *Yes, and I read it*. Because it was likely to be of some importance, was it not? *I have crossed with Captain Smith before and he has handed me messages which have been of no importance at all*, countered Ismay. Were there other passengers present? *There were*. Did you read the message to them? *I did not*. Did you say anything to the passengers about it? *I spoke to two passengers in the afternoon. At that time I did not speak to anybody*.

Lord Mersey and Sir Rufus Isaacs then tried to ascertain at what time Ismay had been given the message. He now said it was before lunch, but at the US inquiry, they recalled, Ismay had told Senator Smith that he did not know if it was before or after lunch. 'I suggest to you that what you said in America was accurate, that you were not certain whether it was in the afternoon or immediately before lunch?' *I am practically certain it was before lunch*, Ismay recalled. The fact that he was heading off to lunch would explain why he had put the

message in his pocket and then forgotten about it; the reason had
been that he was otherwise occupied rather than that he did not want
the ship to reduce speed. He had lunched alone, and later that afternoon he encountered Mrs Thayer and Mrs Ryerson on the deck. *I
cannot recollect what I said*, he told the inquiry of the last time he had
properly seen Marian Thayer, *I think I read part of the message to them
about the ice and the derelict – not the derelict, but that steamer that was
broken down; short of coal she was.* He was speaking rapidly.

Did you understand from that telegram that the ice which was
reported was in your track? *I did not.* Did you attribute any importance at all to the ice report? *I did not. No special importance at all.*
Then why, asked Isaacs, do you think the Captain handed you the
message? *As a matter of information, I take it,* Ismay said, as though
information was generally understood to be a thing of no importance. Information, countered Isaacs, of what? *About,* replied Ismay,
the contents of the message. Isaacs was exasperated. When it came to
Marconigrams, Ismay was Yamsi again, back through the looking-
glass. The Marconigram, he suggested, referred to nothing but
itself; the message contained information about *the contents of the
message.*

The question of the Marconigram was returned to during the
afternoon by other interrogators. Did you before that particular
Sunday, asked Thomas Scanlan, who was determined to catch Ismay
in his own trap, know what was the practice with regard to
Marconigrams received by the officers on the ship relating to the
navigation of the ship? *I believe*, replied Ismay, *the practice was to put
them up in the chart room for the officers.* Was not the Marconigram
from the *Baltic* essentially a message affecting navigation? *Yes.* Then
will you say why, under those circumstances, with that knowledge,
you put that Marconigram into your pocket? *Because,* Ismay
responded patiently, *it was given to me, as I believe now, just before
lunchtime, and I went down and had it in my pocket.* According to
Ismay, the reason he had put the Marconigram in his pocket was
because he *had it in my pocket.* And you suggest that you put it in
your pocket simply in a fit of absent-mindedness? *Yes, entirely.* And

you still retained it in your pocket until it was asked for by Captain Smith late in the evening? *Ten minutes past seven, I think it was, he asked me for it.* That is to say, it had been in your possession for something like five hours? *Yes, I should think so.* (Ismay will have dressed for dinner, in which case he would have removed the message from the jacket he was wearing during the day in order to put it into the pocket of his black evening jacket.) Did you not ask the Captain, wondered Scanlan, when you returned the message whether the ship would come within the latitude and longitude indicated? *I did not,* said Ismay. And the Captain said nothing to you about it? *He did not.* As far as Ismay was concerned, the Captain had handed him the message for no reason, and it said nothing.

On a voyage, wrote Filson Young in *Titanic*, two lives are lived side by side, 'the life of the passengers and the life of the ship'. Ismay insisted that he lived the life of the passenger, but he also secretly lived the life of the ship. The question of his status on the *Titanic*, of whether or not he was an ordinary passenger, rested on the message from the *Baltic*. Why had Captain Smith given him the Marconigram if not to raise with him the issue of speed? Why had Ismay then treated it as a private letter rather than a public document? Why did he keep it back from the officers' chart room but reveal its contents to two female passengers, as though the life of the ship had been any concern of theirs? Ismay claimed later that day, as he had claimed during the US inquiry, that he did not *understand latitude and longitude* and therefore had no idea what the message signified. Then the Marconigram, Isaacs put it to him, was unintelligible to you, was it not? *It was unintelligible to me,* Ismay conceded, *as far as latitude and longitude were concerned.* Ismay had read the message, he said, but not its meaning; he did not speak the language of ships and the sea: *the Marconi message did not convey any meaning to me as to the exact position of ice.* When he looked at the ocean, Ismay saw nothing but a waste of water while the Captain and officers saw the ship's position on the globe as clearly as if it had been a star in the galaxy. But you understood, Isaacs pressed, that you would be in the ice region reported during the night? *Yes, I understood that we would get up to the ice that night.* So how, Isaacs wondered, did Ismay know that they were approaching a region of ice if not from the Marconigram? Because, Ismay quietly explained, Dr O'Loughlin had told him so over dinner. This is what the two men had been discussing during that last supper in the dining room. But Ismay also told the inquiry that he *knew we were approaching the region of ice* because of *this Marconi message.* So was it not, Lord Mersey then asked Ismay to confirm, because of the Marconigram that you expected to come into the region of ice that night? *Oh yes,* replied Ismay, *it was because of the Marconigram.* But, reasoned Mersey, if latitude and longitude are the things which tell you where the ice is and you do not under-stand latitude and longitude, I am quite at a loss to understand why it is you say you came to the conclusion that you would be in the ice region because of the Marconigram?

Ismay's story came down, as ever, to a problem of meaning or even to the meaning of meaning. 'He might have meant something else,' explained Sir Rufus Isaacs to the baffled Lord Mersey, 'but he does mean on account of the Marconigram.' What he meant, Ismay explained, was that he knew they were reaching the ice region on account of the Marconigram *and* the doctor's comment: it was the combination, he said, of *the two together.* 'Then what you mean is this, concluded Isaacs, that you presumed the *Baltic* had sent you a message, without knowing whether it was right or wrong, apprising you of ice in your track? That is what you mean to say? *Yes.* That is what Ismay meant to say.

To the British inquiry, the meaning of the Marconigram lay not simply in its contents but in its being passed from Captain Smith to Ismay. Its meaning lay in its being inside Ismay's pocket rather than on the board of the officers' chart room. But the meaning of the Marconigram could also be found in the peculiar status of the message itself. The *Titanic* was not a lonely boat upon a lonely ocean; the sea was an invisible network of countless tracks and crossroads, and so too was the air above her masts. The transatlantic cable, laid along the ocean floor by Brunel's *Great Eastern*, had allowed telegraph messages to be sent across the Empire, but Guglielmo Marconi's recent invention of wireless messaging allowed ships for the first time to communicate with one another without using flags and flames. It was this, above all else, that interested Filson Young in the story of the *Titanic*; Young, who would later be influential in the development of BBC radio broadcasts, was quick to see the potential of new media technologies. Marconi had called into being a previously unknown universe of waves and currents; he turned a great silence into a thousand twangling instruments which carried, as Young put it, 'whispers, questions, summonses, narratives'. The dots and dashes of Morse code dancing with the speed of light along threads of aerial wire allowed every ship within a thousand miles to spend their days in idle gossip. Through the headphones of Marconi's operators, steamers swapped their stories, describing where they were, the weather around them, and the strange sights they had seen.

The scientific community had remained sceptical about the claims Marconi initially made for his invention – Thomas Edison believed that Marconi's ambition to send wireless over water using electromagnetic charge was impossible – but in 1906 he set up a large transmitting station at Clifden in Connemara and a regular wireless telegraphy service began. By 1908 the messages sent from his apparatus were known as Marconigrams; in 1909 he received the Nobel Prize; in 1911 he launched the journal *The Marconigraph,* and in March 1912 the official offices of the Marconi International Marine Communication Company opened in London. Marconi was regarded as a hero, someone who had made the sea as solid and stable as a suburban street. Should anything go wrong on a voyage, the wireless system would come to the rescue. The year Marconi became a Nobel laureate, White Star's *Republic* collided with another ship, was ripped to the water line and plunged into total darkness. Her Marconi operator succeeded in alerting a number of rescue vessels before directing the *Baltic* – the same ship whose ice warning the *Titanic* later received – to the wreck through thick fog. This was the

first instance of wireless saving lives: Ismay had no reason to doubt
that this could happen again.

But there was something unnerving as well as reassuring about
Marconi's invention. He had turned the solidity of the word into a
fluttering, flying, melting thing which entered the material world as
if through a rent in the mist. Marconigrams were born in transition,
they belonged nowhere, they moved between realms. They were
really no more stable than a butterfly being buffeted about on the
current of a breeze. 'When it has anything to say to you,' wrote Filson
Young, it 'whispers in your ear in whining, insinuating confidence.
And you must listen attentively and with a mind concentrated on
your own business if you are to receive from it what concerns you,
and reject what does not; for it is not always the loudest whisper that
is the most important.' The *Burma*, one of the ships to be contacted
by the *Titanic*, did not hear her whisper until one hour and twenty-
eight minutes after she had sunk. The *Titanic*'s last words, it seemed,
had simply hung about the atmosphere before finally, posthumously,
making themselves heard.

Incidents such as this led Marconi to become increasingly
convinced that sounds never die but remain suspended on their
threads, growing weaker by the minute until we can no longer hear
them. With the right apparatus, he believed, it would be possible to
catch in his net the sound of Christ delivering the Sermon on the
Mount. In Rabelais's *Gargantua and Pantagruel*, written four centuries earlier, the giant Pantagruel realises that the air above their ship
is filled with frozen words and 'some of us . . . cupped our palms
behind our ears' to hear them, before plucking them in fistfuls and
throwing them onto the deck where, 'like sweets of many colours',
they melted away into sounds which no longer had a meaning: '*Hing,
hing, hing, hing, hisse*', said the words as they dissolved; '*hickory,
dickory, dock; brededing, bededac, frr, frrr, frrr, bou, bou, bou, bou, bou,
bou bou. Ong, ong, ong, ong, ouououong; Gog, magog.*' The sailors
were hearing the clashing, clanging, neighing and wailing of a battle
long since fought; there must be an 'inquiry', Pantagruel suggests,
'into whether this may be the very place where Words unfreeze'.

Rabelais's characters try to preserve these sounds in oil, to wrap them in straw and pickle them in jars, but when Marconi's telegraph operators snatched their sounds from space and stamped them onto paper they were no more able than Pantagruel to ensure their readability.[6]

Ismay's story buzzes with new technology: he receives telephone calls from Mrs Thayer, he communicates by telegraph with the White Star Line in New York, and Marconigrams flit across the sea, continually going astray. 'Every message so far', reported the *Northern Echo* thirty-six hours after the *Titanic* disaster, 'has been by wireless, and has come from some steamship officer or land station officer who has picked up others emitted from the keyboard of the *Carpathia*, the *Olympic*, the *Parisian*, the *Virginian*, the *Baltic*, or the *Californian*, and intercepted in their mysterious passage through the air.' Marconigrams were harbingers of information – 'bald statements of fact', as Lightoller put it – but also carriers of misinformation; they seldom arrived on time, no one knew who sent them or to whom they were addressed, they could be intercepted by anyone or not read at all, and they seemed never to be in the right place. The origin of the Marconigrams which claimed that the *Titanic* was being towed to Halifax with all her passengers on board was never found, but they were probably the result of Chinese whispers, words in the ether gone awry. It was pure chance that the *Carpathia* had picked up the *Titanic*'s distress signal at all; the first message Ismay sent following the rescue, in which Franklin was told that the *Titanic* had sunk and that more information was on its way, did not arrive for two days and the further information promised would never be sent. Ismay's later 'Yamsi' messages failed to answer Franklin's urgent questions, and it was unclear anyway whether they had come from Ismay or from Lightoller, who told the US inquiry that while he should be held 'responsible' for their contents he had neither written nor delivered them. Four days after the sinking, New York's White Star Line office reported, 'Many messages have been sent to the *Carpathia*, but we could get no response to our inquiries. I have received absolutely no details of the actual loss of the vessel, and we

know nothing about what has happened, except such scrappy information as has been contained in the few authentic wireless messages
received and already made in public. I have had a code message
from Mr Bruce Ismay, but it relates to business and throws no light
whatever on the tragedy.'

The *Titanic* had either received, or intercepted, eighteen ice warnings during her final weekend, several of which never made their way
out of the Operators' Room. The Marconi operators, Jack Phillips
and Harold Bride, were employed by the Marconi Company and not
the White Star Line; neither members of the crew nor under the
command of Captain Smith, they were unknown figures on the ship,
housed as they were in their own quarters where they divided their
time between dispatching lucrative private messages (such as 'Hello
Boy. Dining with you tonight in spirit, heart with you always. Best
love, Girl', sent three hours before the *Titanic* foundered) and receiving navigational reports whose levels of importance they decided
upon themselves. Ismay had said that the *practice* was to *put them up
in the chart room for the officers*, but Marconigrams were lawless
missives which carried the risk of never being read even if they did
reach their destination. Asked by Senator Smith whether he had seen
the ice reports on the board of the chart room, Lightoller said that he
had not, 'because I did not look'. As far as Captain Smith was
concerned, he had been commanding ships for decades without the
use of wireless technology; it was experience that counted, not
Marconigrams. The trouble with Marconigrams, as Ismay learned to
his cost, is that it was impossible to tell whether they were important
or unimportant.

The message from the *Baltic* whispered to Ismay from a fork in the
air, it had been hammered into solid form like the voice of a dead
man being rapped out on a table-top. Marconi operators were spirit
mediums and the words they transcribed were always already dissolv-
ing. The Marconigram was evidence of communication without
presence; it proved the possibility of connecting with other worlds. It
was a pure example of the kind of 'contact' that fascinated Mrs
Thayer, and when Ismay saw her on the deck it was not the message

but the medium he was showing her. He was flashing around his toy, as a man one hundred years later might demonstrate his new phone. What struck Ismay now, as that final afternoon was frozen for examination by the British inquiry, is that he had held the future in the palm of his hand; the Marconigram had been a psychic forewarning of the ship's fate.

───────

Sir Rufus Isaacs put the issue of the Marconigram aside for the moment to question Ismay on the speed of the ship. How fast was the *Titanic* going on her final day? *I really have no absolute knowledge myself,* replied Ismay, *as to the number of revolutions. I believe she was going seventy-five on the Sunday.* But really Mr Ismay, responded Isaacs with evident exasperation, if you will just search your recollection a little. Remember that this question of speed interested you very materially. You, as Managing Director of the Company, were interested in the speed of the vessel? *Naturally,* Ismay conceded. *Our intention,* he then said, had been *if the weather was suitable on the Monday or Tuesday, to run the ship at her full speed of seventy-eight revolutions.* With whom, wondered Isaacs, would you discuss this question of driving her at full speed on the Monday or Tuesday? Presumably, the court assumed, this decision would have been made in conversation with the Captain, but Ismay had insisted that he had not talked to the Captain at all during the voyage. *The only man I spoke to in regard to it,* he said, *was the Chief Engineer in my room when the ship was in Queenstown.* Will you explain that? asked Isaacs. It is not quite clear why you should discuss the question in Queenstown? The *Titanic* had stopped for passengers first at Cherbourg and then at Queenstown in Ireland, before beginning her journey to New York. *The reason why we discussed it at Queenstown was this,* explained Ismay at length. *Mr Bell came into my room; I wanted to know how much coal we had on board the ship, because the ship left after the coal strike was on, and he told me. I then spoke to him about the ship and I said it is not possible for the ship to arrive in New York on Tuesday. Therefore there is no object in pushing her. We will arrive there at 5 o'clock on Wednesday morning, and it will be good landing for the passengers in New York, and we shall also be able to economise our coal. We did not want to burn any more coal than we needed.*

'Never mind about that,' said Isaacs, brushing aside the issue of the ship's coal supply, 'that does not answer the question I was putting to you . . . the question I am putting to you is this, when was it that you discussed putting her at full speed on the Monday or the Tuesday?' The discussion about the ship's speed had been *at the same time* as the discussion about coal, said Ismay, and had taken place in his cabin on the Thursday afternoon. 'You have not told us about that,' Isaacs noted, as Ismay fell further down the well. Ismay's only other mention of Joseph Bell, the Chief Engineer who died that night, had been in the account of his actions following the collision, when the two men had met on the main staircase and Ismay had asked him the extent of the damage to the ship. This earlier scene, in which Ismay and Bell sat whispering head to head in his cabin at Queenstown, had fortunately escaped the notice of Senator Smith and its revelation now was damning evidence of Ismay's role as, in Conrad's words, a 'double captain'. Bell, one year older than Ismay, was from Maryport, the home town of Thomas Ismay and the place to which Ismay Senior would remain loyal throughout his life, making a point of employing Maryport residents on the White Star liners. Joseph Bell and Bruce Ismay had the same roots; of all the people on board the *Titanic*, Bell was the man with whom Ismay shared the most.

'Then you did know on the Sunday morning,' Isaacs continued, 'that in the ordinary course of things between then and the Monday evening you might be increasing your speed to full speed?' *I knew*, Ismay said, *if the weather was suitable either on the Monday or the Tuesday the vessel would go at full speed for a few hours.* In order for the ship to reach her maximum level of revolutions, further boilers would have to be lit, which implied that the *Titanic* had been gathering speed on the Sunday in preparation for her trials on Monday.

When Ismay was asked whether the Captain was aware that the ship was to increase her speed, he answered, simply, *No*.

'Now, Mr Ismay,' said the Irish Thomas Scanlan, who had taken over the questioning. 'I want to ask you this question: What right had you, as an ordinary passenger, to decide the speed the ship was to go at, without consultation with the Captain?' Ismay did not have time to answer, because Lord Mersey intervened.

'Well, I can answer that – none; you are asking him something which is quite obvious; he has no right to dictate what the speed is to be.'

'But he may,' said Scanlan, 'as a super captain.'

Ismay apparently suppressed a smile of embarrassment as nervous laughter ran through the hall.

'What sort of person', inquired Mersey, 'is a "super captain"?'

'I will tell you as I conceive it, My Lord,' said Scanlan. 'It is a man like Mr Ismay who can say to the Chief Engineer of a ship what speed the ship is to be run at.'

A 'super captain', like Nietzsche's 'superman', is beyond accountability.

Was Ismay a super captain, a double captain, or a double agent, living both the life of the ship and the life of the passenger? In America, he had spoken of *our intention* to run the ship at full speed on the Monday or Tuesday but Senator Smith had not asked to whom the 'our' referred. Ismay, it was now pointed out, repeatedly used the word 'our' in relation to the *Titanic*, and it was drawn to his attention that during the US inquiry he had said that *we were working her up* to full speed. 'You see, you use the personal pronoun, "we".' *I could not*, Ismay drily replied, *have said I was gradually working her up*. But, it was suggested, Ismay could have said 'the Captain' rather than 'we', if it was indeed the Captain whom Ismay was talking about.

On 18 April, a statement had appeared in the *Daily Mail* by Alexander Carlisle, the brother-in-law of Lord Pirrie and former general manager of Harland & Wolff. In the original plans for the *Titanic*, Carlisle revealed, provision had been made for having four lifeboats per davit. Questioned today about Carlisle's claims, Ismay denied having ever seen *any such design*. Nor did he *know that anybody connected with the White Star Line saw such a design*. Moments later he conceded that, *I saw the design I have no doubt; I saw the design with the rest of the ship*, after which he once again insisted that *I tell you I have never seen any such design*. Have you ever heard of it before? asked Lord Mersey. *No, I have not*, Ismay confirmed.

'I take it,' said Scanlan, 'that this is what you say, that you have no recollection of seeing the design at all.' Ismay replied that this indeed was the case. 'Or the fitting up of the boats at all?' *Oh, yes*, he said that he had seen the design for the fitting up of the boats. 'You did see it?' asked Scanlan. *Oh, yes*, repeated Ismay.

Lord Mersey tried to clarify things. 'Have you ever until today heard that there was a design for the *Titanic* by which she was to be provided with forty lifeboats?' *No, My Lord*, Ismay said.

Ismay's meanings were slipping in and out of focus, and Scanlan moved the questions on.

––––––––

Thomas Scanlan's interrogation had cornered Ismay, but it was the Welsh MP, Clement Edwards, who had him pinned wriggling to the wall. Edwards focused on Ismay's 'moral duty' on board the ship and his right to survive the *Titanic* at all. 'You', he asked, 'were one of those responsible for determining the number of lifeboats?' Ismay agreed. 'Did you know that there were some hundreds of people on that ship who must go down with her?' Ismay said that he had known this. 'Has it occurred to you,' asked Edwards, 'that, apart from the Captain, you, as the responsible managing director, deciding the number of boats, owed your life to every other person on that ship?' *No*, said Ismay firmly, *it has not*. If, Edwards reasoned, Ismay had been helping to load the women and children into the lifeboats, then why had he not searched for further women and children on other decks before jumping into Collapsible C himself? A man in his position should surely show himself prepared to take some level of personal risk. Ismay's not having moved from the starboard side looked suspiciously as if he was ensuring his own survival rather than taking responsibility for the ship's community of 'other' passengers.

Ismay replied in what had by now become his mantra: *I was standing by the boat, I helped everybody into the boat that was there, as the boat was being lowered away, I got in.*

That does not answer the question, said Edwards. You had been taking a responsible part, according to the evidence and according to your own admission, in directing the filling of the boats? Ismay disagreed; he had not been taking a 'responsible part', he had simply *been helping to put the women and children into the boats as they came forward.* Edwards sighed. I am afraid, he said, we are a little at cross-purposes. Is it not the fact that you were calling out 'Women and children first', and helping them in? *Yes*, Ismay conceded, *it is.* Then, reasoned Edwards, is it not the fact that you were giving directions as to women and children getting in? *I was*, Ismay said, *helping the women and children in.* Please, demanded Edwards, answer my question. Is it not the fact that you were giving directions in helping them? *I was calling*, Ismay explained, *for the women and children to come in.* Calling for the women and children and then helping them in was not the same thing, Ismay suggested, as giving directions. What I am putting to you, said Edwards, is this: that if you could take an active part at that stage, why did you not continue the active part and give instruction, or go yourself to other decks, or round the other side of that deck, to see if there were other people who might find a place in your boat? *I presumed*, said Ismay, *that there were people down below who were sending the people up.* But, Edwards drove in his rapier, you knew there were hundreds who had not come up . . . That is your answer, that you presumed that there were people down below sending them up? *Yes.* And does it follow from that that you presumed that everybody was coming up who wanted to come up? *I knew*, Ismay finally admitted, *that everybody could not be up.* There was a silence. Then, Edwards continued slowly, I do not quite see the point of the answer? *Everybody*, Ismay snapped, *that was on the deck got into that boat.*

Lord Mersey then intervened.

'Your point, Mr Edwards, as I understand is this: that, having regard to his position, it was his duty to remain upon that ship until she went to the bottom. That is your point?'

'Frankly,' said Edwards, 'that is so; I do not flinch from it a little bit.' *There were no more passengers to get into that boat*, Ismay then repeated for the umpteenth time; *the boat was actually being lowered away.*

———

The next day, in anticipation of Ismay's reappearance at the stand, the Drill Hall was filled to its capacity. 'Fashionably dressed dames', the *Daily Sketch* reported, 'levelled lorgnettes and opera glasses,' through which they saw 'a quietly dressed and rather youthful man of unassuming mien step up to take the oath'. But anyone looking for a thrill was too late: they had missed the main event and the concluding session of Ismay's part in the inquiry, which was dominated by the safe questioning of White Star's own lawyer, Sir Robert Finlay, stayed firmly on technical lines. Only once did Ismay's 'quiet, stolid' voice, the *Telegraph* reported, verge on the 'dramatic', and this was when he was again asked about the circumstances under which he left the *Titanic*. Where were the other passengers? *I can only assume,* he said, *that the passengers had gone to the after part of the ship.* He was no longer leaving behind him an empty ship; Ismay's story was filling up.

At the back of the hall, amongst the myriad faces gazing at the witness stand, was a most discreet and understanding man. In his dramatic illustration of Ismay's interrogation, 'J. Bruce Ismay Before the British Titanic Inquiry', which took up a double-page spread of the popular illustrated news journal, *The Sphere,* Fortunino Matania sees him as no one has seen him before. Ismay is an outcast marooned on an island, cut off from the rest of mankind by a sea of facts. He is a figure whose very simplicity complicates matters. The Drill Hall is a forest of shining pates, white collars and frock coats – all seen from the back – conferring, scribbling, and straining to hear. This is Conrad's British Board of Trade, taking 'its dear old bald head' out for a moment from under its wing. The room is stuffed with words, talk crossing over more talk like radio waves, and the balconies overflow with feathered hats and craning necks. On ground level, in the centre of the first bench, Sir Rufus Isaacs fires his questions at the diminutive figure who is positioned far away on the stand with his back to the model of his now dead ship, one hand in his pocket, the fingers of his other hand touching the table. The only person to be represented in full length, Ismay advances straight at you, his feet treading lightly on the ground. Two weeks ago, he was indistinguishable from these featureless men; he belonged to the same realm of boardrooms and rapidly ascending power. Now he has the look of a man who has been to the edge of the world and back. To Matania, Ismay is an enigma, locked so tightly within that he seems like a missing person. He has massive presence and no presence at all, he is both the smallest and the largest man in the room, and we see him as clearly or as unclearly as we see ourselves. 'At sea, you know,' Marlow explains in *Chance* in words which could have been written to accompany Matania's image, 'there is no gallery. You hear no tormenting echoes of your own littleness there.'

The inquiry, Ismay told Mrs Thayer, had been 'the most trying ordeal to go through. They had me on the stand for seven hours and I was not in a fit condition either mentally or physically to give evidence.' His appearance was followed by that of his friend and business partner, Harold Sanderson, and later by Sir Ernest Shackleton, the Arctic explorer. Asked about speed, Shackleton replied that the case of the *Titanic* opened up the 'very wide question of relationship between owners and captains'. Captains, Shackleton believed, act 'under the instructions of their owners . . . when the owner is on board, you go'.

The proceedings dragged out until the end of the month, after which counsel considered their findings. During a speech which lasted three days – longer than any of Marlow's monologues – Rufus

Isaacs concluded that 'We are left, I must say, in some difficulty in understanding what actually took place between Mr Ismay and Captain Smith.' Despite the acumen and ferocity of his interrogation and the impact in the courtroom of Scanlan's term 'super captain', Isaacs did not 'think that there is any evidence that Mr Ismay interfered. The evidence that we have got all tends the other way.' It was as though the Solicitor-General had heard nothing of the proceedings at all. 'But,' he continued, 'as your lordship has pointed out, and it must be pretty obvious, the showing of the telegram to Mr Ismay was not such an act as would have been performed by the Captain to an ordinary passenger.' Ismay was rather, as the *Telegraph* put it, a 'super passenger'. Sir Robert Finlay then referred to a suggestion made during the proceedings that when Ismay read the Marconigram from the *Baltic* he 'ought to have said to the Captain to "Go Slow"'. 'Sometimes,' Finlay sorrowfully concluded, 'it seems Ismay was to blame for interfering and at others to blame for not interfering.'[7]

The commission's report was presented on 30 July. It was less florid than the narrative produced by Senator Smith, more British in its restraint. Their verdict was, Ismay told Mrs Thayer, along the lines he expected and no doubt others expected it too, but for many people Mersey's concluding remarks would come as a surprise. The British inquiry into the wreck of the *Titanic* decided that no one was responsible for the death of 1,500 people. There was some tut-tutting about more lifeboats clearly being needed, which in future should be manned by seamen rather than passengers, but otherwise the wreck was the result of excessive speed and extraordinary climatic circumstances. Lightoller's rhetoric had won the day. If Captain Smith had indeed – and it was, Lord Mersey insisted, 'pure surmise' – questioned Ismay on the issue of navigation, he was given a rap around the knuckles, for 'no one can suppose for a moment that the Captain did not know quite well that the whole responsibility of the navigation of the ship was upon him and that he had no business to take orders from anybody else'. As for Ismay, it was agreed that his status on the ship was 'in a category all its own' but that his 'very presence . . . had an effect on the navigation . . . even though he never said a single word'. Here the British inquiry endorsed, indeed took its script from, the US Senate inquiry. As for the question of whether Ismay had performed his 'moral duty' – if, as Mersey put it, the 'discharge of the moral duty of Mr Ismay was relevant' – he was cleared of blame. 'Mr Ismay,' said Lord Mersey, 'after rendering assistance to many passengers, found "C" collapsible, the last boat on the starboard side, actually being lowered. No other people were there at the time. There was room for him and he jumped in. Had he not jumped in he would merely have added one more life, namely, his own, to the number of those lost.'

Lord Mersey approached Ismay's situation as though he had been making silent inquiries into his own case. Exonerating Ismay, he exonerated himself; 'Nobody', as Captain Marlow said, 'is good enough.'

———

As he and Florence dined on the evening of 5 June, at 15 Hill Street, Ismay's ordeal in the Drill Hall now over, a mile down the road in the Albert Hall a 'One Hundred Years Ago' ball was taking place. It was billed as the party of the season; the great dome was transformed into Brighton's Royal Pavilion during the time of the Regent, and 4,000 people dressed in the fashions of 1812 came to celebrate Wellington's victory at Waterloo. 'The spirit of Beau Brummel was abroad,' the papers reported, 'and it was reflected in the magnificent costumes and dresses which were worn . . . the scene after midnight was brilliant in the extreme. It was easily the most fashionable and the most brilliant society function of the season. The floor was a fluttering, jingling, dazzling maelstrom.' The ball was a nostalgic reminder of simpler, slower, happier days, but it also echoed the maelstrom of the last moments of the *Titanic*, where the combination of music, dressing gowns and dinner suits made the scene something like, as one passenger said, 'a fancy dress ball in Dante's Hell'. 'Would life have been then more pleasant', the *Daily Mirror* wondered the morning after the London party, 'than it is today as we know it? Some enterprising spirit might next year organise a Futurist Ball which shall transport us to 2012 . . . In 2012 we shall dread that war in the air is coming. If you are unhappy today, 1812 would have suited you no better, and no better than 2012 which, on notepaper, will look like a telephone number misplaced.'

8

Ismay's Unrest

They said I got away in a boat
And humbled me at the inquiry. I tell you
  I sank as far that night as any
Hero. As I sat shivering on the dark water
  I turned to ice to hear my costly
Life go thundering down in a pandemonium of
  Prams, pianos, sideboards, winches,
Boilers bursting and shredded ragtime. Now I hide
  In a lonely house behind the sea
Where the tide leaves broken toys and hat-boxes
  Silently at my door. The showers of
April, flowers of May mean nothing to me, nor the
  Late light of June, when my gardener
Describes to strangers how the old man stays in bed
  On seaward mornings after nights of
Wind, takes his cocaine and will see no-one. Then it is
  I drown again with all those dim
Lost faces I never understood. My poor soul
  Screams out in the starlight, heart
Breaks loose and rolls down like a stone.
  Include me in your lamentations.

Derek Mahon, 'After the *Titanic*'[1]

Cashla, meaning 'twisting creek' or inlet from the sea, is a Gaelic-speaking part of Connemara in County Galway. Ismay called the place by its anglicised name of Costelloe, and it was here that in 1913 he began what Walter Lord describes in *A Night to Remember* the life of 'a virtual recluse'. Ismay's fall, which was reported in newspapers across America, had become the stuff of legend. A feature in Oklahoma's *Times Democrat* in 1914 described him as living 'among the missing' in the 'bleakest' place on earth. 'Look where he hides in misery and shame. Day after day he must hear them, the shrieks of drowning men crying down the wind.' In 1915, Utah's *Ogden Standard* called Ismay 'The World's Exile': 'In one of the wildest spots on the west coast of Ireland where the silence is so heavy that one hardly dares to speak, lives in exile a man who until a few years ago had a high social standing in New York, had wealth and enjoyed all the pleasures and sports that his rank and financial standing afforded.' Now, it was reported, local children chanted 'coward coward coward' when they passed his gate.

Costelloe Lodge, as Ismay's house was called, lay in an oozing, seeping, percolating moonscape of bogs, loughs and boulders, gazed down upon by the pale blue form of the Twelve Pins Mountains. 'It is awfully wild and away from everybody,' he told Mrs Thayer after his first visit; 'I will enjoy the place.' The treeless limestone terrain, scooped out, scraped smooth, grooved and rent by ice-age glaciers, was inhabited by goats rather than people. The River Cashla, carving its course through the moor, cut through the ten acres of Ismay's garden, enclosing the house on two sides. In nearby Clifden, the masts at the Marconi station picked up and passed on the stories that blew in from the sea. Like the house in Liverpool in which he had grown up, the windows of the Lodge looked out to the Atlantic.

To live in isolation is an appropriate penance for someone who has committed a breach of faith with the community of mankind. The symbolism of Ismay's fate seems perfect: a seaman in exile from the sea, he hides on a primitive island which is also the womb of his ship, the flame of his honour keeping its vigil inside him like the last candle burning in a desecrated church. This is his Promethean punishment: Ismay the Titan bound forever to a rock, peregrines and ravens circling his head, the screams of the dead in his ears, frozen words hanging on the air, the empty sky and the empty ocean shimmering together as far as the vanishing horizon.

———

But that is not the story of the second half of Ismay's life. The reality was harder than the public imagined it to be: Ismay simply carried on living, keeping out of the way and out of his own way. He did not retire to Costelloe Lodge and hide behind the sea; he visited Ireland for long summer breaks because the Cashla had the best trout- and salmon-fishing in all of Ireland. It was rumoured that every fish he caught on his hook reminded him of the drowned from the *Titanic*, but the truth is that Ismay found in fishing some of the forgetfulness he sought. The angler, says Isaac Walton, is 'free from the unsupportable burthen of an accusing, tormenting Conscience: a misery that no-one can bear . . .' Ismay was respected by the locals, to whom he gave employment and by whom he was called 'Your Honour', and his children, grandchildren and close friends would join him in Costelloe for memorable holidays. After her husband's death, Florence continued to come here with the family. 'For the first time,' wrote Ismay's granddaughter, Pauline Matarasso of her childhood visits to Connemara, 'in this place of wilderness and wet I found a reality that was better than books. House and garden, large but not grand, stood close by an inlet where the peat boats came from the Aran Isles with their red-brown sails tied up to unload the turf that was the islanders' only income.' Ismay, who had always been private, now became more private; as Pauline Matarasso puts it, he reverted to type. 'When he laid aside his public persona in 1913 he stood stripped to basics – a man so emotionally inhibited and so narrow in his interests as to be inapt for normal family and social life. Reclusion is the choice of those for whom the chronic pain of isolation seems preferable to the agony of rebuff. In this sense, he had long been a recluse.'[2] The catastrophe had confirmed for him that the true danger was neither icebergs nor instincts; it was other people.

As arranged, on 30 June 1913, Ismay handed over the presidency of the IMM and the chairmanship of the White Star Line to Harold Sanderson. When the Great War began, the White Star liners were turned into armed cruisers and the Ismay family left Sandheys in Liverpool to live in the London house at Hill Street. White Star had a good war: the *Olympic* – 'Old Reliable' as she became known – survived after ramming and sinking a submarine, and when America joined the Allies, the *Baltic* was proud to bring over the first US troops. After the *Titanic*, Ismay had set up a pension fund for maritime widows to which he donated £10,000. In 1919, he put forward a further £25,000 to begin a National Mercantile Marine Fund in appreciation of the 'splendid and gallant' bravery of the men in the British Mercantile Marine, with preference for grants and pensions to be given to those who had served on

Liverpool-built ships. Ismay gave generously to charities for the rest
of his life.

Meanwhile, compensation claims were being presented. In
October 1913, when an Irish farmer sued for the loss of his son he was
represented by Ismay's adversary, Thomas Scanlan, who interrogated
Lightoller for a second time, on this occasion before a jury. The
White Star Line was now found guilty of negligence, the jury
concluding that the danger to the ship was foreseen and the speed
reckless. White Star appealed but the verdict was upheld; the conclu-
sions drawn by the British Board of Trade inquiry had been made to
look ridiculous. In America, at the instigation of the *Titanic* Survivors'
Committee, suit was commenced against White Star on behalf of the
steerage passengers. Those survivors who had not been called as
witnesses at the Senate inquiry now provided depositions to show
that Ismay had been running the ship. Elizabeth Lines, a first-class
passenger returning from Paris for her son's graduation ceremony,
testified that after lunch on the day before the collision she had over-
heard a two-hour conversation between Ismay and Captain Smith.
The two men were taking coffee in the lounge, a few tables away
from her own, and Mrs Lines had her back to where they were sitting.
'At first I did not pay any attention to what they were saying, they
were simply talking and I was occupied, and then my attention was
arrested by hearing the day's run discussed, which I already knew had
been a very good one in the preceding twenty-four hours, and I
heard Mr Ismay – it was Mr Ismay who did all the talking – I heard
him give the length of the run and I heard him say "Well, we did
better today than we did yesterday, we made a better run today than
we did yesterday, we will make a better run tomorrow. Things are
working smoothly, the machinery is bearing the test, the boilers are
working well." They went on discussing it, and then I heard him
make the statement, "We will beat the *Olympic* and get in to New
York on Tuesday."' Those exact words, Mrs Lines said, 'fixed themselves' on her mind. Captain Smith had said nothing throughout and
Ismay's tone had been 'dictatorial'. The owner 'asked no questions, he
made assertions, he made statements. I did not hear him defer to

Captain Smith at all.' Elizabeth Lines admitted that she had only
seen Ismay once in her life, twenty years before when he lived in New
York, but insisted that she knew 'for certain' the voice belonged to
him. He was the same man who ate his meals 'at the Captain's table
on the Captain's right'.[3]

Emily Ryerson now provided the testimony she had withheld
from Senator Smith during the US inquiry. She had been walking
on the deck with Marian Thayer when Ismay appeared and
'produced from his pocket a telegram'. 'We are in amongst the
icebergs,' he said, and will 'start some extra boilers tonight'. Being
in a state of 'mental distress' at the time and finding his company
boring, she had barely listened. Her statement was less a memory
than 'the record of an impression left on my mind', but Ismay's
manner was 'that of one in authority and the owner of the ship and
that what he said was law. If this can be of service to anyone I do
not wish to be silent or seem to be protecting him.'[4] Soon after Mrs
Ryerson made this statement, Marian Thayer stopped replying to
Ismay's letters.

On 24 March 1914, before the King's Bench Division of His
Majesty's High Court of Justice, Ismay once again told his story, this
time to George Betts, the New York attorney for the American
claimants. As Ismay would not return to America, Betts had come
to him. He was questioned about his position as a 'conspicuous
passenger', about his understanding of the Marconigram from the
Baltic, his discussion with Joseph Bell at Queenstown concerning
the ship's speed trials, and his communication with Captain Smith
following the collision. Ismay's rhetorical device was now vagueness,
his manner that of someone grown tired of the facts. He casually
contradicted himself, and to ensure that what he said today bore
some relation to what he had said at the inquiries two years before,
he requested reminders of his previous replies to the same
questions.

When asked with whom he dined on the night of the accident,
he said 'Dr O'Loughlin, I think it was.' Of the time at which the
two men ate, he replied 'somewhere about half past 7 . . . I think it

was'. The captain was dining 'with some ladies, I think Mrs Widener'; as to the presence at the Wideners' dinner party of Mrs Thayer, 'I do not know whether she was there or not'. Asked to confirm whether the day on which he was handed the Marconigram from the *Baltic* was Sunday 14 April, Ismay said that it was a Sunday but that he 'did not know the date'; asked whether he had shown the Marconigram to anybody else on the ship, he said that he 'did not remember'. About whether he remembered 'speaking on the afternoon of Sunday on deck to Mrs Ryerson and showing her the message', he remembered speaking only 'to Mrs Thayer'. The Marconigram, Ismay said, did not make 'very much impression on me with regard to the ice', and as for the question of speed, he had 'never really considered the matter'. He described going onto the bridge in his pyjamas and talking to the Captain, and said that it was a 'long time' after this that he realised the ship would sink. Questioned as to whether he was familiar with the practice of putting ice warnings on the board of the chart room, Ismay said that 'it would be a natural thing to do'. He swore that he had never dined at the Captain's table – this had been the privilege of the Thayer family – but on one occasion, he now 'believed', the Captain may have dined with him. He had 'never', Ismay said, sat 'in the ship's lounge discussing the passage with Captain Smith', and he had 'never' told the Captain that 'we will beat the *Olympic* and get into New York on Tuesday'.

The claimants' case against the White Star Line finally commenced in the United States Court in New York on 22 June 1915. Claims ranged from $50, by Eugene Daly, for a set of bagpipes, to $177,352.75 by Mrs Cardeza – the richest passenger on the ship – for the loss of fourteen trunks, three crates, a jewel box and four suitcases. Among the other items passengers claimed for were a signed photograph of Garibaldi, an Arab costume, a marmalade machine and 10lb of tea. With the help of a host of new witnesses, including Karl Behr and Jack Thayer, George Betts from the law firm of Hunt, Hill and Betts argued that Ismay had behaved on board as though he were a super captain, while the defence contended that he was travelling as an ordinary passenger. On 28 July 1916, the judge signed the decree ending all *Titanic* suits and White Star settled out of court, paying out $664,000 in compensation for loss of life and luggage, the price of a life being $30,000. It was a tacit admission of guilt.

In 1925, the IMM, which had never been profitable, got rid of its foreign flag holdings; passenger airlines crossing the Atlantic would soon be making steamships a thing of the past. Two years later White Star was bought by Lord Kylsant, chairman of the expanding Royal Mail Steam Packet Company. It was Kylsant, known as Lord of the Seven Seas, who finally sank the White Star Line; in 1931 he was found guilty of filing fraudulent financial reports, stripped of his title and sentenced to twelve months in Wormwood Scrubs. The government stepped in to save the company, forcing them into a merger with their great rival, Cunard. In 1933, Thomas Ismay's life's work became known as Cunard White Star Limited, and the Liverpool and London offices closed down.

———

As someone who feared chaos, Ismay now arranged for himself a life of routine. Easter was spent golfing at the Gleneagles Hotel in Scotland and summers were spent in Ireland. For the rest of the year he would catch the train every Sunday afternoon from Euston to Liverpool Lime Street, draw the blinds of his private compartment and eat the cold supper he had brought with him. He would then stay three nights in the city's North Western Railway Hotel and attend to his business meetings, returning to London on Wednesday evening. He was still director of the London and North-Western Railway (a forerunner of British Rail), of the London and Globe Insurance Company, the Sea Insurance Company and the Pacific Loan and Investment Company, and he was chairman of the Asiatic Steamship Company and the Liverpool and London Steamship Protection and Indemnity Association, whose minutes show that the *Titanic* was mentioned on a weekly basis in relation to insurance claims. If Ismay went alone to a concert at St George's Hall, as he preferred to do, he booked two seats and placed his hat and coat on the second one. Florence arranged her bridge parties for the half of the week that her husband was away, and when the couple hosted the occasional dinner – for never more than eight people – Bruce would always be given his own plate of cold turkey.

In 1922, Costelloe Lodge was destroyed in an arson attack by the IRA – these were the years of the War of Independence, when big houses were everywhere going up in flames – but Ismay, undeterred, had it rebuilt brick by brick, taking great interest in the project. The new lodge was designed by Sir Edwin Lutyens and the garden laid out by Gertrude Jekyll. In 1927, Ismay's second daughter, Evelyn, married Harold Sanderson's son, Basil, uniting the two families in a way that would have pleased Thomas Ismay. Evelyn later completed the convergence of the twain by giving birth to Ismay-Sanderson twins.

Evelyn's daughter, Pauline Matarasso, remembers the 'solemnity' of her grandparents' Hill Street home. 'Anyone arriving at the house was met by a triumvirate: two footmen, one as dark as the other was fair, would be standing on the steps to either side of the door, flanking the heavier figure of the butler, poised to take the name of the visitor.' She does not recall Ismay speaking, 'though I suppose he must have given us some form of greeting.'⁵ Adding to the austerity of the atmosphere was the silence around the subject of the *Titanic*. To the horror of the dinner table, one child once asked his grandfather whether he had ever been in a shipwreck, while another, proud to be able to impart such adult information, told Ismay that he had read in the newspapers about a train crash in which 256 people had died. 'How do you know 256 people died?' Ismay retorted. 'Were you there? Did you count them?' It was not until Ismay's grandchildren were older that they would hear the story of the *Titanic* and learn that it was never to be mentioned. 'Truth', as Pauline Matarasso puts it, 'was of little interest to the grown-ups gathered round my childhood. They were in the possession of the facts, which cleared my grandfather of wrongdoing. This concentration on the facts was very calming.

Without a moral dimension, without words like hubris, competition, guilt, greed, hedonism, the event was drained of emotion like a stuck pig of blood. It was their way of surviving; they could cope with a corpse. My grandfather Ismay was a corpse himself.'[6]

Ismay spent his last summer in Ireland in 1936, after which acute pains in his right leg led to an amputation below the knee, which took place in his bedroom in Hill Street. Now housebound in London and no longer able to enjoy his outdoor pursuits, he withdrew further into himself. On 27 May of the same year, Cunard White Star's first ship, the *Queen Mary*, left Southampton on her maiden voyage – 'Say,' quipped an American passenger as she boarded the liner, 'when does this place reach New York?' – and on 20 September 1937 the *Titanic*'s twin, the *Olympic*, was taken to Inverkeithing in Scotland to be dismantled. Three weeks later, Ismay suffered a stroke which left him unable to speak or see. He died aged seventy-four on 17 October 1937, the year of George VI's coronation. Following his funeral at St Paul's in Knightsbridge, Ismay was cremated and his ashes placed in a shady plot at Putney Vale Cemetery in south London. The four sides of his tomb were engraved by Alfred Gerrard, Head of Sculpture at the Slade School of Fine Art, with a handsome fleet of sailing ships. The stone contains lines from James 3:4:

> *Behold also the ships which though they be so great and are driven of fierce winds yet are turned about with a very small helm whithersoever the governor listeth.*

In the garden of Costelloe Lodge, Florence had the following words carved into a stone:

> *In memory of Bruce Ismay, who spent many happy hours here 1913–36. He loved all wild and solitary places, where we taste the pleasure of believing that what we see is boundless as we wish our souls to be.*

After the inquiry into the case of the *Patna* finds Jim guilty of 'abandoning in the moment of danger the lives and property 'confided to' his 'charge', he is stripped of his seaman's papers. He stays on in the eastern seaports, limping from job to job like a bird with a broken wing, hiding his identity, unable to endure reference to the scandal, and disappearing whenever it is mentioned. It is impossible, he realises, to lay the ghost of a fact; 'there is always that bally thing at the back of my head.' Jim wants a 'clean slate' and with Marlow's help he makes 'the second desperate leap of his life', into Patusan, a remote island backwater in the South Seas consisting of bandits, maidens, and thirty miles of forest and surf. In the novel's second half, Jim is treated as a leader and called by the natives '*Tuan*' or 'Lord' Jim. Just as he once gazed at the serenity of the world from the security of the *Patna*, he now looks 'with an owner's eye at the peace of the evening, at the river, at the houses, at the everlasting life of the forests, at the life of the old mankind, at the secrets of the land'. Defending his village from attack by an English marauder, Jim dies like a medieval knight.

'The division of the book into two parts', Conrad admitted, was its 'plague spot'.[7] Readers and critics have agreed; *Lord Jim* is less a book divided into two than a self-divided book, a book which breaks in the middle and, like *Wuthering Heights* and *David Copperfield*, it has become a novel remembered for its first part alone. Conrad was successfully able to imagine Jim's failure but he failed to imagine his success; when his account of Jim's life in Patusan begins, Marlow relaxes his grip on the young man's consciousness and the narrative loses its psychological tautness. Conrad would have been wiser to say goodbye to Jim as he embarked on his island life, as he said goodbye to Leggatt in 'The Secret Sharer' when he slipped off the ship to swim to Koh-ring, a 'free man . . . striking out for his new destiny'. But he was too involved in Jim's story, he needed too badly see Jim wrest control of his fate, and in order for Jim to win back his honour Conrad had to change the genre of his book. What Albert Guerard described as 'perhaps the first major novel built on a true intuitive understanding of sympathetic identification as a psychic process', now became a boy's adventure story of the kind that Jim had always dreamt of inhabiting.

Jim eventually dies for honour, but has he redeemed himself? Has he, in his island life, faced or shirked his crime? While Conrad made Jim's ending too easy, Marlow is always drawn to the equivocal and it is the absence of clarity in the young man's situation that interests him. Despite Marlow's endless production of words, Jim becomes less and less clear as the tale progresses. It is only, Marlow realises, 'when we try to grapple with another man's intimate need that we perceive how incomprehensible, wavering and misty are the beings that share with us the sight of the stars and the warmth of the sun'. By the close of *Lord Jim*, the book's subject is no longer the encounter between the self we believe ourselves to be and the self unknown, lying in wait like a snake beneath a stone. Marlow now wonders whether atonement is possible at all, whether a slate can ever be clean. 'A clean slate, did he say? As if the initial word of each our destiny were not graven in imperishable characters upon the face of a rock.'

The second half of lives, as well as novels, can also be of limited interest. Few people much mind what Wordsworth had to say after he married, and the story of Florence Nightingale ends once she extinguishes her lamp, her final forty years being consigned to a footnote. Conrad used the half of his life spent on land to translate into romance the monotony of the half he spent at sea, and what is curious about Ismay's life is that having jumped from the *Titanic* into Conrad's world, a place in which we live out our time as convicts, self-deceived and unpitied, his story was quickly given a Conradian second half. Ismay was imagined as an outcast on an island even as he attended to his daily rhythm, walking the streets of London in a suit with a rolled umbrella, taking the train to Liverpool, feeding the pigeons in the park. The man whose speech was once described by the press as a 'luminous fog' was now constructed by them as a man in a mist. And that, it was assumed of Ismay, is the end. He passes away under a cloud, inscrutable at heart, forgotten, unforgiven, and excessively romantic.

By the time of Ismay's death, the *Titanic* was no longer the most horrific event of the twentieth century and the crisis of one man's loss of honour had, for the moment, been forgotten. The tone of his obituaries was gentle, forgiving; the *London Evening News* said that although 'the *Titanic* episode was written on his heart, it will not be his epigraph'. The *New York Times* stated that he had been a 'passenger' on the ship and had died 'without making any further public statement on the *Titanic* or his conduct than that which he told the Senate Committee and Lord Mersey's Board of Trade investigations'. The *Journal of Commerce* remarked that he was 'one of the greatest captains of the shipping industry . . . having once reached a decision [he was] absolutely immovable', added to which he possessed 'an extraordinary memory'. On the 'personal side', however, 'Mr Bruce Ismay was, to most people, an enigma'. In America, the *New York Tribune* noted of Ismay's life that 'the parallel with the tale of Conrad's *Lord Jim* will occur to most of us'. The parallel had occurred to no one, and if it had ever occurred to Conrad – as it must have done – he did not say so. Conrad turned his face away from the human side of the *Titanic*. He knew what ships did to men, he had written enough about the codes of fidelity and community on which both he and Ismay had been raised. Nor would he have considered the mirroring of Jim's and Ismay's stories such an extraordinary coincidence. Nothing about the sea surprised Conrad: so long as there are ships to sink, men will jump from them.

'Jim's desertion of his ship', the *New York Tribune*'s obituary of Ismay continued, 'was, to be sure, in direct violation of his duty . . . but it too had its explanation that men accepted. The point is that with a sensitive soul the explanation, however convincing, is overshadowed by the necessity to explain, and from that there seems to be no release short of the grave.' Ismay felt the necessity to excuse rather than explain himself and while Jim, too, makes excuses – 'I told you I jumped; but I tell you they were too much for any man. It was their doing as plainly as if they had reached up with a boathook and pulled me over' – he is also compelled, as Ismay was not, by the desire to understand his actions, or at least for others to understand them. Ismay had no knowledge of the release that comes from elucidation, and only the slightest idea – from his brief communication with Mrs Thayer – of the healing power of empathy. He was happy to appear as a man without moral content because, unlike Jim, he had not dreamed of becoming a hero. Ismay had never thought so much of himself.

According to Wilton Oldham in *The Ismay Line*, the only time Ismay ever spoke of his departure from the ship was in a conversation with his sister-in-law, Constance, who was the sister of Bruce's wife as well as the wife of his brother. Described by her great-niece as a 'bookworm, animal-lover, traveller, rebel, and in her old age [a] dispenser of wisdom',[8] Constance, who thought Florence's moratorium on talking about the *Titanic* a mistake, was another of Ismay's secret sharers. The reason, he revealed to her, that he had jumped was not because the boat was there and the deck was empty, but because he been ordered to do so by Chief Officer Wilde, who said that Ismay's evidence would be needed at the subsequent inquiry. Ismay left the ship knowing that the purpose of his survival was to represent the Captain, the crew and the company, to be the witness of the night, to contain, explain and make sense of it all to the bewildered world. In exchange for his continued life, he must carry into the history books the story of the *Titanic*. Why did he keep this a secret? Both Albert Weikman, the *Titanic's* barber, and Lightoller had told Senator Smith that Ismay had been thrown into the lifeboat by Wilde, but Ismay had always denied that this was the case. After hearing Lightoller's defence of Ismay in Washington, Lawrence Beesley wrote in the *New York Times* on 29 April 1912 that 'bundling Ismay into a boat . . . seems a very natural act for an officer of the line to perform toward the head of the line'. Endorsing, in his conversation with Constance, this version of events, Ismay now suggested that the great burden of his life was not that he had jumped from a sinking ship teeming with lives, but that he made a Faustian pact which he then failed to keep. Ismay decided, as he sat in the lifeboat, that he would rather be cast as a coward than drag behind him the ball and chain of the narrative. His life had been granted in order that he face the inquiry, but it was Lightoller instead who took on this role and played it for all it was worth.

Fidelity to the notion of honour was a concept that would dominate the lives of Ismay's two sons-in-law. Brigadier General Ronald Cheape, the husband of Ismay's eldest daughter, Margaret, was a soldier possessed, as Pauline Matarasso puts it, more of 'bone-headed courage than tactical brilliance'. Cheape, known as 'the General', refused to speak to one of his sons – named Bruce – who had spent four years in Germany as a prisoner of war because, as far as he was concerned, the young man was a 'hands-upper'. Never happier than during a war, the General lived like a Viking on his Scottish estate and expected his guests to do so too; even lavatory paper – which the Picts had done without – was considered by him effeminate. He was, in the words of one of his nephews, 'the last of the Titans'. According to Ronald Cheape's obituarist, 'There are many people with physical courage, fewer with moral courage. General Cheape's outstanding characteristic was the combination of both in the highest degree.'

Basil Sanderson, the husband of Evelyn, was more refined in his understanding of courage. Having fought on the Somme and won an MC, Basil wrote in an army notebook an essay containing his thoughts on 'Fear'. 'There is no such thing as courage,' he begins, 'at least in the way most people understand the word.' Many men overcome fear not because they are brave but because they have an even greater fear of scorn. Because he was 'born a coward' in a culture of heroes and hero-worship, young Basil spent his boyhood preparing to defend himself against the sudden emergence of fear, jumping into the deep end of swimming pools because he was afraid of water, taking up boxing so that people would think he was brave, banging at the door to enlist in 1914 because 'the one thing that has always petrified me is the thought of war'. These, he says, are signs not of courage but of 'moral cowardice', which is a condition that arises in a man when 'fear of the opinion of the rest of mankind is more terrible than that awful dread which is gnawing at him inside'. Moral cowardice was the reason that Basil Sanderson 'went down into the biggest and longest battle there has ever been and stayed in the line at my own request throughout the whole of it. I was terrified to think that people might suspect I was afraid . . . And oh, the fights and battles which I used to have within myself!'⁹ It is impossible to read these pages without thinking of Ismay, the closest friend of Basil Sanderson's father, who had thought himself impervious to fear until it reared its hydra head.

––––––––

'Why', Marlow wondered of his interest in Jim, 'I longed to go grubbing into the deplorable details of an occurrence which, after all, concerned me no more than as a member of an obscure body of men held together by . . . fidelity to a certain standard of conduct, I can't explain. You may call it unhealthy curiosity if you like; but I have a distinct notion I wished to find something.' Would Marlow have found anything in the deplorable details of Ismay's story? The tragedy for Ismay was not that his ship hit an iceberg, but that he was unable, as Conrad puts it, to find 'fit words for his meaning'. Rather than confide in a discreet, understanding man with an interest in extracting the 'felt, subjective experience behind the objective, outward facts', Ismay watched as his story was constructed from a collage of contradictory witness reports, hasty journalistic reactions and a drama staged by Senator William Alden Smith. Without the presence of a Marlow, the chaos of the night, the horror of it all, remained for Ismay chaos and horror, an experience too great to absorb. Had Marlow been at either of the inquiries into the *Titanic*, he would have known immediately, instinctively, of what Ismay was made. And had he then taken Ismay off to some quiet corner of New York's Waldorf-Astoria or Willard's Hotel in Washington and listened to him circle his story – *I got in as she was being lowered away . . . Because there was room in the boat . . . She was being lowered away . . . I felt the ship was going down and I got*

into the boat . . . I did not get into the boat until after they had begun
to lower it away . . . My back was turned to her . . . I did not wish to
see her go down – he would have experienced one of his moments of
awakening, 'when we see, hear, understand, ever so much – everything – in a flash', and when he sees a person 'as though I had never
seen him before'.

Then Marlow would stretch out his legs after dinner on the deck
of some barque, light his cigar, fill his glass, and tell Ismay's tale to an
audience of men who also follow the sea. First he would paint on his
dark background the details so essential to the myth of the *Titanic*:
there would be a ship the size of a cathedral, her monstrous birth in
the Belfast shipyard; the decision to limit her lifeboats so as not to
clutter the decks; her doomed beauty; the cheering, the pride, the
jubilation as she slides down her cradle to taste her first drop of
water; the ice warnings; the Captain driving her on and on, the
moonless sky, the sudden appearance of the berg, rising 500 feet out
of the water; the misplacement of the binoculars; the boys in the
crow's nest ringing the bell; the order to turn 'hard-a-starboard'; the
opening up of the ship like a tin of sardines; the torrential rush of
water; the sleeping passengers; the dutiful crew; the Captain losing
control; the band playing ragtime; the *Californian*, with her engines
off only eight miles away; the steerage passengers trapped down
below; the half-filled boats dropping into the water; the men in their
dinner jackets going down like gentlemen; the man who dressed as a
woman to get a place in a lifeboat; the wives who chose to die with
their husbands; the other wives in the lifeboats refusing to save their
husbands; the strange unnatural calm of the conditions; the refusal
of the passengers to take in what was happening until the very end;
the small town's-worth of people who died that night. Marlow would
linger over the many different languages spoken in the steerage
compartments, the four Chinese sailors of Collapsible C, and he
would save for his finest canvas the splendour of the *Titanic*'s final
dive and the death-music that followed. But at the heart of his story
would be Ismay's jump and his subsequent battle with his moral
identity, because for Marlow 'the ship we serve is the moral symbol

of our life' and nothing can be said with certainty about a man until he has been 'tested' by his ship.

Marlow and Conrad parted company two weeks before the *Titanic* set sail. His departure is as mysterious as his arrival, but Marlow had probably seen enough, analysed too much, sat for too long upon too many decks holding audiences spellbound with too many soliloquies, and Conrad said goodbye to him on the night of 25 March 1912 when he finished *Chance: A Tale in Two Parts*, whose subject is coincidence. 'The last words were written at 3.10 a.m.,' he told the American collector of his manuscripts, John Quinn, 'just as my working lamp began to burn dimly and the fire in the grate to turn black . . . I went out and walked in the drive for half an hour. It was raining and the night was still very black.'[10] He was happy with what he had produced but 'as to what will happen to it when it is launched, I am much less confident'.[11] That spring, *Chance* was serialised in the *New York Herald,* making Marlow's appearance in New York converge with the attacks on Ismay in the American press.

The *Titanic* is a tale of the convergence of art with nature, but nothing converges so much in its telling as fact and fiction. Even for those survivors who watched from their lifeboats as the ship went down, it was an imagined as well as a real experience and many later drew on Dante and Virgil in their descriptions of the night. Conrad, too, preferred to base his yarns on recorded events; the origin of *Lord Jim* was, after all, the affair of the *Jeddah* and Jim himself was modelled on the *Jeddah*'s Chief Mate, Augustine Podmore Williams. Conrad used Jim as a way of reflecting on the conduct of Williams, and one of the effects of using Jim's story to reflect on Ismay's is to see how characters who live in fiction have more appeal than those who do not, although even Marlow admits to becoming 'thoroughly sick' of Jim's 'vapourings'. Ismay is less sympathetic than Jim, just as an evening spent with Hamlet at a hotel bar would be less engaging than an evening spent watching him perform his indecision on the stage, and Emma Bovary would become a bore were she to telephone us every day. As with a mirror, the distance afforded by art adds depth of vision; art increases our capacity for sympathy. It reveals, as Oscar Wilde puts it in 'The Decay of Lying', 'nature's lack of design, her curious crudities, her extraordinary monotony, her absolutely unfinished condition'. It is only when we place Ismay's crude, monotonous, absolutely unfinished narrative next to that of *Lord Jim* that his form begins to thicken, his blood to flow and his consciousness to take on an essential extra layer. The voluble Jim, who digs deep into his own experience, is the underside of the taciturn Ismay who, having glimpsed the depths, sticks to the surface of his story throughout.

Conrad and Ismay converge as well: born five years apart, Ismay is all size and splendour while Conrad, a solitary exile, is a creature of great experience and terrifying intensity who comes to us by sea from heaven knows where. Even to his friends, Conrad was a stranger drifting through. Like the *Titanic* itself, the lives of both men broke in half somewhere near the middle; each spent the first part on water and the second part on land, recovering. Their love of ships pushed them to the limit. Each collapsed beneath the weight of his experience, each found solace from the sea on an island, and each jumped – twice.

In the mailroom of the *Titanic* was a package containing the manuscript of 'Karain: A Memory', the precursor to *Lord Jim*. Written in 1897 and first published in *Blackwood's*, 'Karain' was later included in Conrad's collection of five stories, *Tales of Unrest*. He had been sending it to John Quinn in New York, whose letter describing as a 'God-send' the US Senate inquiry into the wreck would subsequently inspire Conrad's second article for the *English Review*, 'Certain Aspects of the Admirable Inquiry into the Loss of the *Titanic*'.

Karain is a 'great Bugi dandy', a spotlessly clean, magnificently theatrical Malayan 'adventurer of the sea, outcast, and ruler', whose domain consists of three villages on the narrow plain of an island shaped like a young moon. He befriends the crew of a visiting ship, one of whom, in a prefiguring of Marlow, is the sympathetic narrator of the story. One night Karain swims out to the yawl in a state of terror. 'Not one of us doubted that we were looking at a fugitive . . . He was haggard, as though he had not slept for weeks; he had become lean, as though he had not eaten for days . . . his face showed another kind of fatigue, the tormented weariness, the anger and the fear of a struggle against a thought, an idea – against something that cannot be grappled, that never rests – a shadow, a nothing, unconquerable and immortal, that preys upon life.' In the safety of one of the cabins, Karain reveals that he is an exile on this island; a victim of unrest.

Many years ago, in his own land, the lovely sister of Karain's great friend, Matara, had run away with a red-faced, red-headed Dutch tradesman, bringing dishonour on her family. Karain swears to help Matara avenge himself on the couple, and the two men set out on their journey. After years of scouring the islands in search of the girl, she starts to appear to Karain in waking dreams. 'No one saw her, no one heard her, she was mine only! . . . And she was sad!' Her continued presence 'gave me courage to bear weariness and hardships. Those were times of pain, and she soothed me . . . She was all mine and no one could see her.' Karain, who talks to the vision in the dark, murmurs to her one night, 'you shall not die'. When he and Matara at last find the house where the girl and the Dutchman are living, Matara hands the gun to Karain and whispers 'Let her die by my hand. You take aim at that fat swine there. Let him see me strike my shame off the face of the earth – and then . . . You are my friend – kill with a sure shot.' Karain takes the gun and then sees the tender eyes of the 'consoler of sleepless nights, of weary days; the companion of troubled years . . . Had I not promised that she should not die? . . . her voice murmured, whispered above and around me, "Who shall be thy companion, who shall console thee if I die?"' Karain hears himself shout to her to run and the girl leaps; Matara too leaps and runs towards his sister; Karain fires the gun and kills Matara instantly. 'The sunshine fell on my back colder than the running water', and he walks away into a forest 'which was very sombre and very sad'. Since that day Karain has been 'hunted by his thought along the very limit of human endurance'. The girl stopped appearing to him – 'Never! Never once! She had forgotten' – and instead he is pursued by the spirit of Matara, who 'runs side by side without footsteps, whispering, whispering old words – whispering into my ear in his old voice'.

The crew listen to Karain's tale in silence and try to console him: 'Every one of us, you'll admit, has been haunted by some woman,' they say. To protect the chief from the whispering shade of his murdered friend, they give him a talisman in the form of a coin depicting the head of Queen Victoria. It is unclear whether they are helping him to face or to shirk his crime, but Karain, the narrator concludes, 'had known remorse and power, and no man can demand more from life'.

Marian Thayer, the woman who constantly occupied Ismay's thoughts and to whom he had talked 'all the time' in the year following the wreck, died on 14 April 1944. It was exactly thirty-two years to the day after the death of her husband on the *Titanic*. The curious timing of Mrs Thayer's death recalls the lines by Louis MacNeice on the death of his grandmother. As a boy in Belfast, MacNeice had 'one shining glimpse of a boat so big it was named *Titanic*':

> *As now for this old tired lady who sails*
> *Toward her own iceberg calm and slow;*
> *We hardly hear the screws, we hardly*
> *Can think her back her four score years.*

> *. . . the day went down*
> *To the sea in a ship, it was grey April,*
> *The daffodils in her garden waited*
> *To make her a wreath, the iceberg waited;*
> *At eight in the evening the ship went down.*

Marian Thayer never remarried and nor did she remove from her dressing table the framed verses Ismay had sent her in 1913. Jack Thayer waited until 1940 to add his own version of events to the sea of survivors' stories. 'No two happenings in the stream of space time are identical,' *The Sinking of the SS Titanic* begins, and 'no two individuals no matter how close they may be together on shipboard have the same description of experience to relate'. For Thayer as for Ismay, Lightoller and many others, the *Titanic* was a tale of sleeping and waking, but with the hindsight of twenty-eight years and two world wars, his perspective had changed. It was not only himself who woke to the sound of a nail scraping along the side of the ship. 'To my mind the world of today awoke April 15th, 1912.'[12]

The world went back to sleep again as far as the *Titanic* was concerned, but interest was reawakened in 1955 by Walter Lord's non-fiction novel, *A Night to Remember,* which two years later was filmed at the Pinewood Studios in Elstree, one mile from the school where Ismay had been so unhappy and Conrad, as a guest of the riotous Sanderson family, had been so happy. Kenneth More played a splendidly game Lightoller, the hero of the drama, and Ismay was portrayed by Frank Lawton as a fool in a pair of striped pyjamas. Lightoller would never see himself immortalised; he died of heart failure in 1952, without getting his own command in the White Star Line or ever receiving from the company 'a word of thanks' for taking on the role of 'whipping boy' at the inquiries. 'It must', he concluded, 'have been a very curious psychology that governed the managers of that magnificent line.'[13] The very brave can be very dangerous and, according to Walter Lord, Lightoller was 'too much the romantic individualist, too likely to say what he thought . . . It was fatal for a White Star Line officer to have been associated with the *Titanic*.'[14]

In 1935, Lightoller had written his memoirs, *Titanic and Other Ships*, which he dedicated to Sylvia, 'My persistent wife, who made me do it'. Here he proves himself a man of many-sided courage, an adventurer addicted to the unpredictability of sea life, to the 'feel of something living under my feet'. It is a tale of cyclones, shipwrecks and desert islands, of albatross bones, giant sea bats, and sailing ships with 'towering tiers of bellying canvas'. It is not until chapter thirty that Lightoller arrives at the maiden voyage of the *Titanic*, but he does not repeat here his claim that following the collision Ismay had ordered the Captain to go 'Slow Ahead' or that Robert Hichens had turned the wheel the wrong way. A storyteller with several versions of the night, each one tailored to a specific audience, the story told by Lightoller to his adoring wife must be regarded in the same light as the many other accounts by survivors of their experience. It is a version of events, to be placed on top of other versions much as the Victorian geneticist, Francis Galton, overlapped photographs of faces so that the individual features vanished and the common characteristics were accentuated. Lightoller's role as Ismay's secret sharer was saved for Sylvia's ears alone, and after her husband's death she carried the story around like a loaded gun. In a letter to Walter Lord sent from America in 1956, Sylvia Lightoller revealed that she was hoping to 'get a call from the CBS for their "I've Got a Secret" TV programme as I have rather a good one about the *Titanic*, so if by chance I am chosen to go to New York I should so much like to meet you and have a chat'. Neither CBS nor Walter Lord took up her offer.[15]

Florence Ismay was not amongst those who wrote to Walter Lord with her account of what 'really happened'. The *Titanic* had ruined the first part of her life and she was determined to enjoy the second, which began after Ismay died, at which point she surrendered her British citizenship and took an oath of repatriation with the United States. This old tired lady was ninety-six when her own ship went down on New Year's Eve 1963, between the end of the Chatterley ban and the Beatles' first LP.

'It is easy to jump in,' J. Bruce Ismay had told Harold Sanderson in 1904, 'but it would be difficult, if not impossible, to climb out.' Ismay never climbed out from the hole into which he had fallen and nor did he achieve the catharsis that traditionally comes with tragedy, but when we see him through Conrad's hooded eyes he has

something of the tragic hero. His destiny lay submerged, riding in
wait, ready to leap. He was an ordinary man caught in extraordinary
circumstances, who behaved in a way which only confirmed his
ordinariness. Ismay is the figure we all fear we might be. He is one
of us.

Memorial stone in the garden of Costelloe Lodge

Advertisement for Joseph Conrad's first essay on the *Titanic*
(*Daily Sketch*, 3 May 1912)

Afterword

What solitary icebergs we are . . .

Virginia Woolf, *The Voyage Out*

In August 1924, one year after returning from New York on board the *Majestic* – his single experience of travelling on a White Star liner – Conrad, the most remarkable of Ismay's secret sharers, died of a heart attack. 'Suddenly,' wrote Virginia Woolf, 'without giving us time to arrange our thoughts or prepare our phrases, our guest has left us; and his withdrawal without farewell or ceremony is in keeping with his mysterious arrival, long years ago, to take up his lodging in this country.'

Conrad produced more words on the *Titanic* than any of his literary contemporaries. Even the press expressed surprise at the silence of the writing community in response to the greatest peace-time shipwreck in history. 'Few of the magazines this month have anything to say regarding the terrible disaster in the North Atlantic which has shocked and saddened the whole world,' noted the *Daily Telegraph* in the summer of 1912 in its round-up of journals such as *Blackwood's*, *Cornhill* and *Strand*. It is only 'Mr Joseph Conrad in the *English Review*', who 'makes some comments on the catastrophe'. W. B. Yeats, the national poet of the country in which the *Titanic* was built, said nothing in public about the event at all; Henry James, who had made the same crossing between England and his native

New York on a dozen occasions, mentions the *Titanic* only once, in a condolence letter to friends of the American artist Francis Millet, who was travelling with White House aide Archie Butt. E. M. Forster, despite his sense in *Howards End* that 'any fate was titanic', remained silent, as did D. H. Lawrence. John Galsworthy attended the inquiry in Washington but apart from comments confided to his diary, he wrote nothing about the experience, not even reporting his impressions to Conrad.

G. K. Chesterton, H. G. Wells, George Bernard Shaw and Sir Arthur Conan Doyle produced newspaper copy when they were asked to do so, but there is no mention of the subject in their private correspondence, and nor did it bleed into their subsequent work. Virginia Woolf, who attended the British Board of Trade inquiry in London, said in a letter to her friend Katherine Cox that 'what I should really like to do now, but must refrain, is a full account of the wreck of the *Titanic*. Do you know it's a fact that ships don't sink at that depth, but remain poised half-way down, and become perfectly flat, so that Mrs Stead is now like a pancake, and her eyes like copper coins.' This image is Woolf's only observation on the wreck: instead of writing a full account of the *Titanic* she wrote *The Voyage Out*, about another journey from which there was no return.

It was the general public who discovered a bottomless capacity for reading and writing about the *Titanic*. Those who had never before penned a line were inspired to produce poems by the yard, some of which were collected in special anthologies; newspapers were inexhaustible in their coverage of the story (the *New York Times* allotted seventy-five pages to the *Titanic* in the first week alone); survivors' accounts were rushed out by the press, and popular journalists put together instant 'biographies' of the ship. The *Titanic* brought out the writer in everyone except those who wrote for a living, most of whom, like Ismay, were apparently struck dumb by the event. The novelists, poets and playwrights of the Edwardian age simply couldn't find the words and what they did write served as a cover for talking about something else, like commerce, chance or class. 'World's Largest Metaphor Hits Iceberg', runs a spoof headline in *The Onion*'s book, *Our Dumb Century* (1999). Even Hardy's memorial poem, 'The Convergence of the Twain', written for the souvenir programme of the Covent Garden benefit matinée for the families of the dead on 4 May, can be read as a poem not about the *Titanic* but about his doomed marriage to Emma Gifford.[1] 'Trust a boat on the high seas,' as Conrad puts it in *Lord Jim*, 'to bring out the Irrational that lurks at the bottom of every thought, sentiment, sensation, emotion.' Conrad's insight could serve as the epigraph for the story of J. Bruce Ismay.

Notes

Documents from the Ismay family archive at the National Maritime Museum in Greenwich are identified by TRNISM, and those from the Lord MacQuitty Collection at the National Maritime Museum by LMQ. The sources for all the quotations from the US Senate and British Board of Trade inquiries into the wreck of the *Titanic* can be found online at www.titanicinquiry.org, and are referred to below as Inquiry Proceedings.

PART I: AT SEA

Chapter 1: Chance

1   A full account of the various versions of the loading of Collapsible C, 'Ismay's Escape: Did he jump or was he pushed?', can be found at: www.paullee.com/titanic/ismaysescape.html

2   John B. Thayer, *The Sinking of the SS Titanic, April 14–15, 1912* (Philadelphia, 1940), p. 20.

3   LMQ/7/2/21.

4   Quoted in Logan Marshall, *The Sinking of the Titanic and Great Sea Disasters* (Charlford, 2008), p. 40.

5   Thayer, pp. 18 and 20.

6   Colonel Archibald Gracie, 'The Truth About the *Titanic*', reprinted in *The Story of the Titanic As Told by Its Survivors, Lawrence Beesley,*

Archibald Gracie, Commander Lightoller, Harold Bride, edited by Jack Winocour (Dover, 1960), p. 259.

7   Charlotte Collyer, *The Semi-Monthly Magazine*, 26 May 1912.

8   Quoted in John Wilson Foster, ed., *Titanic* (Penguin, 1999), p. 83.

9   Violet Jessop, *Titanic Survivor* (Sutton Books, 2007), pp. 152–3.

10   Lawrence Beesley, 'The Loss of the SS *Titanic*', in Winocour, ed., *The Story of the* Titanic *As Told by Its Survivors*, p. 41.

11   Gracie, 'The Truth About the *Titanic*', in ibid., p. 150.

12   Commander Lightoller, '*Titanic*', in ibid., p. 278.

13   François Rabelais, *Gargantua and Pantagruel*, translated by M. A. Screech (Penguin, 2006), pp. 829–30.

14   LMQ/7/2/19.

15   LMQ/7/3/B.

16   *Daily Telegraph*, 22 April 1912.

17   LMQ/7/1/15.

Chapter 2: Luckless Yamsi

1   US Senate Inquiry Proceedings, testimony by Captain Rostron.

2   Marconigram sent from the *Olympic* to the *Carpathia*.

3   *The Times*, 20 April 1912.

4   *Atlantic City Daily Press*, 5 May 1912.

5   US Senate Inquiry proceedings.

6   *Daily Telegraph*, 21 April 1912.

7   TRNISM/3/1.

8   John B. Thayer, *Evening Bulletin*, Thursday 14 April 1932.

9   Lawrence Beesley, 'The Loss of the SS *Titanic*', p. 19.

10   Ibid., p. 30.

11   Ibid., p. 9.

12   Ibid., p. 44.

13   Ibid., p. 43.

14   Ibid., p. 42.

15   Joseph Conrad, *Lord Jim: A Tale*, edited with an introduction by Alan H. Simmons, with Notes and Glossary by J. H. Stape (Penguin Classic, 2007), p. 19. All subsequent quotations from *Lord Jim* come from this edition.

16 Beesley, 'The Loss of the SS *Titanic*', p. 32.

17 *Lord Jim*, p. 17.

18 Lightoller, '*Titanic*', p. 296.

19 Unsigned manuscript dated August 1926. I am grateful to Angus Cheape for drawing my attention to this document.

20 US Senate Inquiry Proceedings, testimony by Lightoller, 19 April 1912.

21 Thayer, *The Sinking of the SS Titanic*, p. 30.

22 Lightoller, '*Titanic*', p. 279.

23 British Inquiry Proceedings, testimony by Alexander Carlisle, Question 21284.

24 *Washington Times*, 17 April 1912.

25 Lightoller, '*Titanic*', p. 275.

26 *Washington Times*, 17 April 1912.

27 Ibid.

28 Ibid.

29 Speech by Senator William Alden Smith, US Senate Inquiry Report, 28 May 1912.

30 LMQ/7/2/37.

31 The bogus messages may have been the result of an eavesdropping amateur wireless operator picking up Marconigrams on two different subjects and combining them. 'Is the *Titanic* safe?' and '*Asian* 300 miles west of *Titanic* and towing oil tanker to Halifax' became 'All *Titanic* passengers safe – towing to Halifax'.

32 US Senate Inquiry Proceedings.

33 Ibid.

34 Wyn Craig Wade, *The* Titanic: *End of a Dream* (Weidenfeld and Nicholson, 1980), p. 97.

35 LMQ/7/1/15.

36 LMQ/7/2/22.

37 Beesley, 'The Loss of the SS *Titanic*', p. 58.

38 Ibid., pp. 81–2.

39 Ibid., p. 82.

40 US Senate Inquiry Report, 28 May 1912.

41 Account by Edith Russell, LMQ/7/2/19.

Chapter 3: Youth

1   TRNISM/7/4.

2   Bram Stoker, 'The World's Greatest Shipbuilding Yard', *The World's Work*, Vol IX, March 1907, p. 360.

3   Jules Verne, *A Floating City* (Routledge, 1876), p. 12.

4   R. A. Fletcher, *Travelling Palaces: Luxury in Passenger Steamships* (Pitman and Sons, 1913), p. 30.

5   Letter from James Ismay to his father, TRNISM/10/1.

6   TRNISM/4/1.

7   TRNISM/10/1.

8   Dudley Parker, *The Man of Principle: A View of John Galsworthy* (Heinemann, 1963), p. 36.

9   Catherine Dupré, *John Galsworthy: A Biography* (Collins, 1976), p. 43.

10 John Eddison, *A History of Elstree School* (1979), p. 49.

11 Quoted in Mark Girouard, *The Return to Camelot: Chivalry and the English Gentleman* (Yale University Press, 1981), p. 13.

12 John Galsworthy, *Another Sheaf* (New York, 1919), p. 92.

13 See Peter Parker, *The Old Lie: The Great War and the Public School Ethos* (Constable, 1987).

14 Wilton Joseph Oldham, *The Ismay Line: The White Star Line and the Ismay Family Story* (Journal of Commerce, 1961), p. 42.

15 John Galsworthy, *The Country House* (Heinemann, 1907), p. 58.

16 Horace Annesley Vachell, *The Hill: A Romance of Friendship* (John Murray, 1905), pp. 24–89.

17 Oldham, *The Ismay Line*, p. xxvii.

18 Thomas Ismay can be compared to the media tycoon, Robert Maxwell, another self-made man who was by turns charming and monstrous. Maxwell ran his home, Headington Hill House, as though he were the ruler of a small province, controlling the decisions and behaviour of his nine children and receiving unconditional devotion from the wife he would ridicule in public. In 1968, he gave an interview in *The Times* where he explained that he did not have good working relations with men because 'they tend to be too independent. Men like to have individuality.' What Maxwell was

looking for, he said, was not a man with a mind of his own but 'an extension of the boss'.

19   Andrew Saint, *Richard Norman Shaw* (New Haven, 1976), p. 261.

20   Ibid.

21   Ibid.

22   Lt Colonel Frank Bustard, *Titanic Commutator*, June 1974.

23   Basil Sanderson, *Ships and Sealing Wax: The Memoirs of Basil, Lord Sanderson of Ayot* (Heinemann, 1967), p. 2.

24   *Truth*, 12 April 1888.

25   Sanderson, *Ships and Sealing Wax*, p. 11.

26   Pauline Matarasso, *A Voyage Closed and Done* (Michael Russell, 2005), p. 17.

27   Ibid.

28   Twenty-seven of Margaret Ismay's pocket diaries are kept in the Ismay archive at the National Maritime Museum.

29   Quoted in Paul Louden-Brown, *The White Star Line: An Illustrated History 1870–1934* (Ship Pictorial, 1991), p. v.

30   US Senate Inquiry Proceedings, 'There is nobody left in the firm except myself. It is practically a dead letter now to all intents and purposes.'

31   Jean Strouse, *Morgan: American Financier* (Random House, 1999), p. x.

32   Ibid., p. 394.

33   Ibid., p. 474.

34   Ibid., p. 467.

35   Albert Ballin to the German Embassy in London, quoted in Strouse, p. 463.

36   LMQ/3/6.

37   Strouse, *Morgan: American Financier*, p. 463.

38   Ibid., p. 477.

39   Alan Frederick Lewis, *The Great Pierpont Morgan* (Gollancz, 1949), p. 282.

40   Oldham, *The Ismay Line*, p. 146.

41   Ibid.

42   Ibid., p. 152.

43   To Sir Clinton Dawkins, quoted in Oldham, *The Ismay Line*, p. 158.

44   *The Times*, 3 December 1903.

45   Oldham, *The Ismay Line*, p. 145.

46   Michael Moss and John R. Hume, *Shipbuilders to the World: 125 Years of Harland & Wolff, Belfast, 1861–1986* (Blackstaff, 1986), p. 92.

47   Oldham, *The Ismay Line*, p. 140.

48   Moss and Hume, *Shipbuilders to the World*, p. 108.

49   Oldham, *The Ismay Line*, p. 161.

50   Filson Young, *Titanic* (Grant Richards, 1912), p. 12.

51   R. A. Fletcher, *Travelling Palaces*, p. 225.

52   Oldham, *The Ismay Line*, p. 171.

53   On the night of the disaster, James Ismay is said to have come out of his coma and said, 'Bruce is in trouble, Bruce is in trouble.'

54   LMQ/1/3.

55   *The Titanic Commutator*, Vol. 11, No. 3, 1987, p. 41.

56   Oldham, *The Ismay Line*, p. 18.

Chapter 4: These Bumble-like Proceedings

1   Robert Hughes, *The Real New York* (Hutchinson, 1905), p. 50.

2   James Remington McCarthy, *Peacock Alley: The Romance of the Waldorf Astoria* (Harper, 1931), p. 163.

3   This is also pointed out by Filson Young, *Titanic*, p. 81.

4   Strouse, *Morgan: American Financier*, p. 643.

5   Lightoller, '*Titanic*', p. 301.

6   *Daily Telegraph*, 22 April 1912.

7   Edward Hungerford, *The Story of the Waldorf Astoria* (The Knickerbocker Press, 1925), p. 131.

8   Stephen Biel, *Down with the Old Canoe: A Cultural History of the Titanic Disaster* (W. W. Norton, 1996), p. 30.

9   Ben Proctor, *William Randolph Hearst, the Later Years* (Oxford University Press, 2007), p. 5.

10   Quoted in Michael Davie, *The Titanic: The Full Story of a Tragedy* (Bodley Head, 1986), p. 156.

11   Ibid.

12   Following the wreck of the *Titanic*, the publishers of *Futility* reissued the book as *The Wreck of the Titan*.

13   Lightoller, '*Titanic*', p. 303.

14   Patrick Stenson, *Lights: The Odyssey of C H Lightoller* (London, 1984), p. 206.

15   Lightoller, '*Titanic*', p. 287.

16   LMQ/7/1/20.

17   Lightoller, '*Titanic*', p. 285.

18   Louise Patten, *Good as Gold* (Quercus, 2010), p. 241.

19   Lightoller, '*Titanic*', p. 277.

20   Ibid., p. 282.

21   Ibid., p. 305.

22   Ibid., p. 288.

23   George Behe, *On Board the RMS* Titanic: *Memories of the Maiden Voyage* (Lulu.com, 2011), p. 335.

24   H. M. Marriot, *The Life and Letters of John Galsworthy* (William Heinemann, 1935), p. 340.

25   Sigmund Freud, *Fragment of an Analysis of Hysteria* in *The Penguin Freud Reader*, ed. Adam Phillips (Penguin, 2006), p. 443.

PART II: ON LAND

Chapter 5: The Convergence of the Twain

1   Conrad, letter to James B. Pinker, 22 April 1912, *The Collected Letters of Joseph Conrad*, edited by Frederick R. Karl and Laurence Davies (Cambridge University Press, 1983–2008), Vol. 5.

2   Conrad, letter to Ted Sanderson, 12 October 1912, *Collected Letters*, Vol. 2.

3   Norman Sherry, *Conrad's Eastern World* (Cambridge University Press, 1966), p. 86.

4   Conrad, letter to Galsworthy, 20 July 1900, *Collected Letters*, Vol. 2.

5   Albert Guerard, *Conrad the Novelist* (Harvard, 1958), p. 142.

6   Conrad, Author's Note, *Youth, Heart of Darkness, The End of the Tether* (J. M. Dent and Sons, 1902), p. vi.

7   Conrad, to John Quinn, 10 May 1912, *Collected Letters*, Vol. 5.

8   Conrad, *Notes on Life and Letters* (J. M. Dent and Sons, 1921).

9   Ibid.

10 Virginia Woolf, 'Joseph Conrad', *Collected Essays* (London, 1968), vol 1:302.

11 Quoted in Zdzislaw Najder, *Joseph Conrad: A Life*, translated by Halina Najder (Camden House, 2007), pp. 31–5.

12 Conrad, *A Personal Record: Some Reminiscences* (J. M. Dent and Sons, 1919), p. 194.

13 Najder, *Joseph Conrad*, p. 41.

14 Ibid., pp. 182–3.

15 Conrad to Agnes Sanderson, 22 April 1896, *Collected Letters*, Vol. 9.

Chapter 6: The Secret Sharer

1   TRNISM 1/1.

2   LMQ/7/2/32.

3   Thayer, *The Sinking of the SS Titanic*, pp. 29–30.

4   The letters from Ismay to Marian Thayer, sections of which have been published in Pauline Matarasso, *A Voyage Closed and Done*, are in private hands.

5   TRNISM/1/1.

6   Wilton Oldham, 'The *Titanic* and the Chairman', *The Titanic Commutator*, Vol. 13, No. 2, 1989, p. 25.

7   Oldham, *The Ismay Line*, pp. 219–20.

8   TRNISM/1/1.

9   Ibid.

10 'The *Titanic* and the Chairman', *The Titanic Commutator*, Vol. 13, No. 3, 1989, p. 46.

11 Oldham, *The Ismay Line*, pp. 220–24.

12 John Masefield, *Manchester Guardian*, 16 October 1912.

13 In 'The *Titanic* and the Chairman', his sequel to *The Ismay Line*, Wilton Oldham describes wanting to be Ismay's secret sharer himself. 'Bruce Ismay would have talked about it [the *Titanic*] with the right person, who had a sympathetic understanding and who was a

complete outsider. I am not praising myself at all but Mrs Bruce Ismay, her sister Mrs Bower Ismay, Bruce Ismay's favourite child, Mrs Margaret Cheape, and Mrs Mary Quirk, a very old friend of the family, have all said that he would have talked to me if I had known him.' (*Titanic Commutator*, Vol. 12, No. 1).

Chapter 7: The Super Captain

1   *Illustrated London News*, 4 May 1912.
2   Lightoller, '*Titanic*', p. 304.
3   Michael Davie, *The Titanic*, p. 173.
4   Filson Young, *Titanic*, p. 189.
5   Lightoller, '*Titanic*', p. 305.
6   Rabelais, *Gargantua and Pantagruel*, p. 830.
7   'The *Titanic* and the Chairman', *Titanic Commutator*, Vol. 13, No. 1, 1989.

Chapter 8: Ismay's Unrest

1   The author is aware that the details of his poem are not in all cases factually correct.
2   Matarasso, *A Voyage Closed and Done*, p. 24.
3   The transcripts of the Limitation of Liability Hearings are available at www.titanicinquiry.org.
4   Emily Ryerson gave her evidence first in a letter sent to her lawyer on 18 April 1913 (LMQ/7/2/22), and then in an affidavit for the Limitation of Liabilities hearings.
5   Matarasso, *A Voyage Closed and Done*, p. 71.
6   Ibid., p. 19.
7   Conrad, letter to Edward Garnet, 12 November 1900, *Joseph Conrad: Life and Letters*, ed. G. Jean-Aubry (New York, 1927), I, pp. 298–9.
8   Matarasso, *A Voyage Closed and Done*, p. 18.
9   'Fear' was written by Basil Sanderson in an army issue notebook. Never shown to his children – who discovered it only after his death – it is now kept with his papers in Trinity College, Oxford.

10   Conrad, letter to John Quinn, 27 March 1912, *Collected Letters*, Vol. 5.

11   Conrad, letter to J. B. Pinker, early April 1912, *Collected Letters*, Vol. 5.

12   Thayer, *The Sinking of the SS Titanic*, p. 9.

13   Lightoller, '*Titanic*', p. 305.

14   From Walter Lord, foreword to Patrick Stenson, *Lights*, p. 8.

15   From Sylvia Lightoller to Walter Lord, LMQ/2/4/153. I am grateful to Louise Patten for drawing my attention to these letters.

Afterword

1   The *Titanic* was a tale of fatal convergence for other writers as well. Unknown to Hardy, on the day that his poem was being read by mourners in London, Helen Candee, who had been on board the *Titanic*, had the account of her experience, 'Sealed Orders', published in the American weekly *Collier's*. Mrs Candee began: 'When all the lands were thrilling with the blossomed month of shower and sun, three widely differing craft spread out upon the sea. One sailed from the New World's city of Towers, plowing east. Another coquetted with three near ports of Europe and then sailed West. The third slipped down unnoticed from the glacial North.' The first was the *Carpathia*, the second the *Titanic*, and the third the iceberg. All three were given 'sealed orders', and the meeting of the 'greatest ship on earth' and the 'sinister' iceberg is described as a 'tryst'. 'Across the starlit sea', the frozen groom awaits his 'virgin' bride: 'it was nearly midnight when she shuddered with horror in the embrace of the northern ice'. The same image had been used sixteen years earlier in a little-known poem called 'A Tryst', by the American writer Celia Thaxter. A 'fair ship' and 'an iceberg pale' are drawn together on a moonless night. The iceberg arrives at the appointed spot, 'Like some imperial creature, moving slow', and the 'stately ship' meanwhile, 'with matchless grace' and 'unconscious of her foe,/ Drew near the trysting place.'

Bibliography

The best sources of information about Ismay and the sinking of the *Titanic* are contemporary newspaper reports. Those consulted for this book are:

Newspapers

Atlanta Constitution
Boston Globe
Daily Graphic
Daily Mirror
Daily News
Daily Sketch
Daily Telegraph
Denver Post
Emporia Weekly Gazette
Frankfurter Zeitung
Glasgow Evening Times
Guernsey Press
Illustrated London News
Jersey Journal
John Bull
Liverpool Daily Post
Lloyd's Weekly News

London Evening News
New York American
New York Evening Post
New York Globe
New York Herald
New York Post
New York Times
New York Tribune
New York World Telegram
Northern Whig
Orleans American
Philadelphia Evening Bulletin
San Francisco Examiner
Sharon Herald
St Louis Post Dispatch
The Chronicle
The Ogden Standard
The Sphere
The Times
The Times Democrat
Uttoxeter Advertiser
Washington Post

Journals

Blackwood's
Christian Science Journal
Christian Science Sentinel
Country Life
Engineering
The English Review
Fairplay
Financier
International Marine Engineering
Journal of Commerce

National Magazine
New York Times Book Review
Pall Mall Gazette
The Critic
The Review of Reviews
The Semi-Monthly Magazine
The Syren
The Titanic Commutator
The World's Work
The Woman's Protest

Witness Accounts

Barratt, Nick, *Lost Voices from the* Titanic: *The Definitive Oral History* (Arrow Books, 2009)

Beesley, Lawrence, *The Loss of the SS Titanic: Its Story and Lessons* (Heinemann, 1912)

Behe, George, *On Board the RMS* Titanic: *Memories of the Maiden Voyage* (Lulu.com, 2011)

Jessop, Violet, *Titanic Survivor,* introduced, edited and annotated by John Maxtone-Graham (Sutton Publishing, 2007)

Thayer, John B., *The Sinking of the SS* Titanic *April 14–15, 1912* (Philadelphia, December 1940)

Winocour, Jack, ed., *The Story of the Titanic as Told By Its Survivors, Lawrence Beesley, Archibald Gracie, Commander Lightoller, Harold Bride* (Dover, 1960)

The Sinking of the SS *Titanic*

Barczewski, Stephanie, *Titanic: A Night Remembered* (Hambledon and London, 2004)

Behe, George, *Titanic: Psychic Forewarnings of a Tragedy* (Patrick Stephens, 1988)

Behe, George and Goss, Michael, *Lost at Sea* (Prometheus Books, 1994)

Booth, John and Coughlan, Sean, *Titanic: Signals of Disaster* (White Star, 1993)

Bryceson, Dave, ed., *The Titanic Disaster, as Reported in the British National Press April–July 1912* (W. W. Norton & Co., 1997)

Davie, Michael, *Titanic: The Full Story of a Tragedy* (The Bodley Head, 1986)

Eaton, John and Haas, Charles A., *Titanic: Triumph and Tragedy: A Chronicle in Words and Pictures* (Patrick Stephens, 1994)

Everett, Marshall, *Wreck and Sinking of the Titanic* (L. H. Walter, 1912)

Gibbs, Philip, *The Deathless Story of the Titanic* (Lloyds of London, 1985)

Heyer, Paul, *Titanic Legacy: Disaster as Media Event and Myth* (Praeger, 1995)

Lord, Walter, *A Night to Remember* (Holt, Rhinehart & Winston, 1955)

——————*The Night Lives On* (Penguin, 1986)

Loss of the Steamship Titanic, British Investigation (7Cs Press, 1975)

Marcus, Geoffrey, *The Maiden Voyage* (Allen and Unwin, 1969)

Marshall, Logan, *Sinking of the Titanic and Great Sea Disasters* (John C. Winston Co., 1912)

Pellegrino, Charles, *Her Name, Titanic: The Untold Story of the Sinking and Finding of the Unsinkable Ship* (McGraw-Hill, 1988)

Report on the Loss of the SS Titanic (Blackstaff Press, 1990)

The Titanic Disaster: Report of the Committee on Commerce, US Senate (7Cs Press, 1975)

Wade, Wyn Craig, *The* Titanic: *End of a Dream* (Weidenfeld and Nicolson, 1980)

Young, Filson, *Titanic* (Grant Richards, 1912)

The White Star Line

Beaumont, J. C. H., *Ships and People* (Geoffrey Bles, 1936)

Cooper, Gary, *The Man Who Sank the Titanic? Life and Times of Captain Edward J. Smith* (Witan Books, 1992)

Eaton, John P. and Haas, Charles, *Falling Star: Misadventures of White Star Line Ships* (Patrick Stephens, 1989)

Fletcher, R. A., *Travelling Palaces: Luxury in Passenger Steamships* (Pitman and Sons, 1913)

Hyslop, Donald, Forsyth, Alastair and Jemima, Sheila, eds, *Titanic Voices: The Story of the White Star Line, the Titanic and Southampton* (Southampton City Council, 1994)

Jefferson, Herbert, *Viscount Pirrie of Belfast* (William Mullan, 1948)

Louden-Brown, Paul, *The White Star Line: An Illustrated History 1870–1934* (Ship Pictorial, 1991)

Maxtone-Graham, John, *The Only Way to Cross* (Patrick Stephens, 1983)

Moss, Michael and Hume, John R., *Shipbuilders to the World: 125 Years of Harland and Wolff, Belfast, 1861–1986* (Blackstaff, 1986)

Oldham, Wilton J., *The Ismay Line: The White Star Line and the Ismay Family Story* (Liverpool, Journal of Commerce, 1961)

Sanderson, Basil, *Ships and Sealing Wax: The Memoirs of Lord Sanderson of Ayot* (Heinemann, 1967)

Conrad

The following works by Conrad are all published by J. M. Dent:

Almayer's Folly (1895)

'Karain' (1897), in *Tales of Unrest* (1908)

Lord Jim: A Tale (1900)

Youth, A Narrative, and Two Other Stories (1902)

The Mirror of the Sea (1906)

'The Secret Sharer', in *Twixt Land and Sea: Tales* (1912)

'Some Reflections, Seamanlike and Otherwise, on the Loss of the *Titanic*' (1912), *Notes on Life and Letters* (1925)

'Certain Aspects of the Admirable Inquiry into the Loss of the *Titanic*' (1912), *Notes on Life and Letters* (1925)

A Personal Record (1912)

Chance: A Tale in Two Parts (1913)

The Shadow-Line: A Confession (1917)

'Ocean Travel' (1923), *Last Essays*, ed. Richard Curle (1926)

'The *Torrens*: A Personal Tribute' (1923), *Last Essays*, ed. Richard Curle (1926)

General

Baker, W. J., *A History of the Marconi Company* (Methuen, 1970)

Barker, Dudley, *The Man of Principle: A View of John Galsworthy* (Heinemann, 1963)

Biel, Steven, *Down with the Old Canoe: A Cultural History of the Titanic Disaster* (W. W. Norton, 1996)

Bloom, Harold, ed., *Marlow* (Major Literary Characters) (Chelsea House, 1992)

Conrad, Borys, *My Father Joseph Conrad* (Calder and Boyars, 1970)

Conrad, Joseph, *Collected Letters of Joseph Conrad*, 9 volumes, edited by Frederick R. Karl and Laurence Davies (Cambridge University Press, 1983–2008)

Dupré, Catherine, *John Galsworthy: A Biography* (Collins, 1976)

Eddleman, Ruth Ellie, *The Function of Marlow in Joseph Conrad's Fiction* (MA thesis, 1962)

Foster, John Wilson, *The Titanic Complex: A Cultural Manifest* (Belcouver Press, 1996)

——————*Age of Titanic: Cross-currents of Anglo-American Culture* (Merlin, 2002)

——————ed., *Titanic* (Penguin, 1999)

Girouard, Mark, *The Return to Camelot: Chivalry and the English Gentleman* (Yale University Press, 1981)

Guerard, Albert J., *Conrad the Novelist* (Harvard University Press, 1958)

Harpham, Geoffrey Galt, *One of Us: The Mastery of Joseph Conrad* (Chicago University Press, 1996)

Hay, Eloise Knapp, '*Lord Jim* and *Le Hamlétisme*', 'L'Epoque Conradienne', 1990, *Société Conradienne Français*

Howells, Richard, *The Myth of the Titanic* (Macmillan, 1999)

Hughes, Robert, *The Real New York* (Hutchinson, 1905)

Hughes, Thomas, *Tom Brown's Schooldays* (Macmillan, 1967)

Hungerford, Edward, *The Story of the Waldorf-Astoria*, New York, 1925.

Jack, Ian, 'Women and Children First', in *The Country Formerly Known as Great Britain: Writings, 1989–2009* (Jonathan Cape, 2009)

Johnson, Bruce M., 'Conrad's "Karain" and *Lord Jim*', *Modern Language Quarterly* Vol. XXIV, No. 1, March 1963

Jones, Max, *The Last Great Quest: Captain Scott's Antarctic Sacrifice* (Oxford University Press, 2003)

Langewiesche, William, *The Outlaw Sea: Crime and Chaos on the World's Oceans* (Granta, 2005)

Lunn, Arnold, *The Harrovians: A Tale of Public School Life* (Methuen, 1913)

Matarasso, Pauline, *A Voyage Closed and Done* (Michael Russell, 2005)

McCarthy, James Remington, *Peacock Alley: The Romance of the Waldorf-Astoria* (Harper, 1931)

Miller, Karl, *Doubles: Studies in Literary History* (Oxford University Press, 1985)

Minchin, J. G., *Our Public Schools: Their Influence on English History* (Swan Sennenschein & Co., 1901)

Najder, Zdzislaw, *Joseph Conrad: A Life,* translated by Halina Najder (Camden House, 2007)

Parker, Peter, *The Old Lie: The Great War and the Public School Ethos* (Constable, 1987)

Patten, Louise, *Good As Gold* (Quercus, 2010)

Proctor, Ben, *William Randolph Hearst, the Later Years* (Oxford University Press, 2007)

Rabelais, François, *Gargantua and Pantagruel,* translated by M. A. Screech (Penguin, 2006)

Ray, Martin, ed., *Joseph Conrad: Interviews and Recollections* (Macmillan, 1990)

Rendell, M. J., 'Harrow: The School of Life', in C. E. Pascoe, *Everyday Life in Our Public Schools* (1881)

Robertson, Morgan, *Futility: The Wreck of the Titan* (7Cs Press, 1974)

Rose, Jonathan, *The Edwardian Temperament 1895–1919* (Ohio University Press, 1986)

Saint, Andrew, *Richard Norman Shaw* (Yale University Press, 1976)

Sanderson, I. C. M., *A History of Elstree School and Three Generations of the Sanderson Family* (privately published, 1978)

Sherry, Norman, *Conrad's Eastern World* (Cambridge University Press, 1966)

Stape, J. H., *The Several Lives of Joseph Conrad* (Doubleday, 2007)

——————'Ideology and Rhetoric in Conrad's Essays on the *Titanic*', *Prose Studies*, Vol. 11, No. 1, May 1988

Stenson, Patrick, *Lights: The Odyssey of C. H. Lightoller* (The Bodley Head, 1984)

Strouse, Jean, *Morgan: American Financier* (Random House, 1999)

Vachell, H. A., *The Hill: A Romance of Friendship* (John Murray, 1905)

Verne, Jules, *A Floating City* (Routledge, 1876)

Waller, P. J., *Democracy and Sectarianism: A Political and Social History of Liverpool, 1868–1939* (Liverpool University Press, 1981)

Watt, Ian, *Essays on Conrad* (Cambridge University Press, 2000)

Wilson, F. B., *Sporting Pie* (Chapman & Hall, 1922)

Woolf, Virginia, 'Joseph Conrad', *The Common Reader* (Hogarth Press, 1925)

Acknowledgements

I would like to thank members of the Ismay family for their help and support, especially Pauline Matarasso whose friendship, insights and intelligence made writing this book an unexpected pleasure. Angus Cheape provided tremendous hospitality and some extraordinary documents, and I am extremely grateful to Malcolm Cheape for allowing me use of family photographs. Thank you also to John Cheape, Pascal Lo, Jim Alderson Smith and Alan and Trudi Sanderson. A chance conversation with Kate Bucknell changed everything, and I owe a particular gratitude to Robert and Polly Maguire.

For sharing their knowledge of the *Titanic* and archives of *Titanic* materials, thanks are due to the experts: John Wilson Foster, Paul Lee and George Behe, whose guidance prevented me from hitting many an iceberg. Louise Patten gave me vital information and a very good tea. I am forever in debt to Conrad's brilliant biographer and editor, John Stape, for the constant flow of reading material sent to me and many fine conversations. For discussing the manuscript at various stages I am grateful – once again – to Pauline Matarasso, and also to Ophelia Field, Paul Keegan and David Miller. Anne Chisholm lent me valuable materials, and Candia McWilliam led me towards others. Also of great help were Alex Towli, Michael McCaughan, Lee Kendall, Ada Wordsworth, Anthony Wilson, Mark Bostridge and Neil Rennie.

Copyright photographic material is reproduced by permission of the following: The Ulster Folk and Transport Museum, English Heritage, Mary Evans Picture Library, Getty Images, the Cheape Family, Angus Cheape, Robert Maguire, Don Lynch, the *Titanic*

Historical Society and the Bettmann Archive. I am grateful to Derek Mahon and The Gallery Press for permission to reproduce 'After the Titanic' from *New Collected Poems* (2011).

Finally, thank you to my agent, Sarah Chalfant at the Wylie Agency, and for the hard work of the wonderful team at Bloomsbury – Kate Holland, Catherine Best, Alexa von Hirschberg, Anna Simpson and especially my editor, Michael Fishwick.

Index

ABOUT THE AUTHOR

Frances Wilson was educated at Oxford University and lectured on nineteenth- and twentieth-century English literature for fifteen years before becoming a full-time writer. Her books include *Literary Seductions: Compulsive Writers and Diverted Readers* and *The Ballad of Dorothy Wordsworth: A Life*, which won the British Academy Rose Mary Crawshay Prize. She reviews widely in the British press and is a fellow of the Royal Society of Literature. She divides her time between London and Normandy.

LIST OF
FIRST CLASS PASSENGERS

ROYAL AND U.S. MAIL

S.S. "TITANIC."

Triple Screw · 46,328 Tons.

FROM SOUTHAMPTON AND CHERBOURG
TO NEW YORK
(VIA QUEENSTOWN)
WEDNESDAY, 10th APRIL, 1912.

WHITE STAR LINE.

"OLYMPIC," (Triple Screw 45,324 Tons.
AND
"TITANIC" Triple Screw 46,328 Tons.
THE LARGEST STEAMERS IN THE WORLD

FIRST CLASS PASSENGER LIST
PER
ROYAL AND U.S. MAIL
S.S. "Titanic,"
FROM SOUTHAMPTON AND CHERBOURG
TO NEW YORK
(VIA QUEENSTOWN),
Wednesday, 10th April, 1912.

Captain, E. J. Smith, R.D. (Commr. R.N.R.)
Surgeon, W. F. N. O'Loughlin.   Purser, H. W. McElroy
Ass't Surgeon, J. E. Simpson.   R. L. Barker
Chief Steward, A. Latimer.

Aitken, Miss Elizabeth Walton
Allison, Mr. H. J.
Allison, Mrs. H. J. and Maid
Allison, Miss
Allison, Master and Nurse
Anderson, Mr. Harry
Andrews, Miss Cornelia I.

Andrews, Mr. Thomas
Appleton, Mrs. E. D.
Artagaveytia, Mr. Ramon
Astor, Colonel J. J. and Manservant
Astor, Mrs. J. J. and Maid
Aubart, Mrs. N. and Maid

Barkworth, Mr. A. H.
Baumann, Mr. J.
Baxter, Mrs. James
Baxter, Mr. Quigg
Beattie, Mr. T.
Beckwith, Mr. R. L.
Beckwith, Mrs. R. L.
Behr, Mr. K. H.
Bishop, Mr. D. H.
Bishop, Mrs. D. H.
Bjornstrom, Mr. H.
Blackwell, Mr. Stephen Weart
Blank, Mr. Henry
Bonnell, Miss Caroline
Bonnell, Miss Lily
Borebank, Mr. J. J.
Bowen, Miss
Bowerman, Miss Elsie
Brady, Mr. John B.

Brandeis, Mr. E.
Brayton, Mr. George
Brewe, Dr. Arthur Jackson
Brown, Mrs. J. J.
Brown, Mrs. J. M.
Bucknell, Mrs. W. and Maid
Butt, Major Archibald W.

Cablehoad, Mr. E. P.
Cardell, Mrs. Churchill
Cardeza, Mrs. J. W. M. and Maid
Cardeza, Mr. T. D. M. and Manservant
Carlson, Mr. Frank
Carran, Mr. F. M.
Carran, Mr. J. P.
Carter, Mr. William E.
Carter, Mrs. William E.
Carter, Miss Lucile
Carter, Master William T.
Case, Mr. Howard B.
Cavendish, Mr. T. W.
Cavendish, Mrs. T. W. and Maid
Chaffee, Mr. Herbert F.
Chaffee, Mrs. Herbert F.
Chambers, Mr. N. C.
Chambers, Mrs. N. C.
Cherry, Miss Gladys
Chevré, Mr. Paul
Chibnall, Mrs. E. M. Bowerman
Chisholm, Mr. Robert
Clark, Mr. Walter M.
Clark, Mrs. Walter M.
Clifford, Mr. George Quincy
Colley, Mr. E. P.
Compton, Mrs. A. T.

Compton, Miss S. R.
Compton, Mr. A. T., Jr.
Cornell, Mrs. R. C.
Crafton, Mr. John B.
Crosby, Mr. Edward G.
Crosby, Mrs. Edward G.
Crosby, Miss Harriett
Cummings, Mr. John Bradley
Cummings, Mrs. John Bradley

Daly, Mr. P. D.
Daniel, Mr. Robert W.
Davidson, Mr. Thornton
Davidson, Mrs. Thornton
de Villiers, Mrs. B.
Dick, Mr. A. A.
Dick, Mrs. A. A.
Dodge, Mr. Washington
Dodge, Mrs. Washington
Dodge, Master Washington
Douglas, Mrs. F. C.
Douglas, Mr. W. D.
Douglas, Mrs. W. D. and Maid
Dulles, Mr. William C.

Earnshaw, Mrs. Boulton
Endres, Miss Caroline
Eustis, Miss E. M.
Evans, Miss E.

Flegenheim, Mrs. A.
Flynn, Mr. J. I.
Foreman, Mr. B. L.
Fortune, Mr. Mark
Fortune, Mrs. Mark
Fortune, Miss Ethel
Fortune, Miss Alice
Fortune, Miss Mabel
Fortune, Mr. Charles
Franklin, Mr. T. P.
Frauenthal, Mr. T. G.
Frauenthal, Dr. Henry
Frauenthal, Mrs. Henry
Frolicher, Miss Marguerite
Futrelle, Mr. J.
Futrelle, Mrs. J.

Gee, Mr. Arthur
Gibson, Mrs. L.
Gibson, Miss D.
Goldenberg, Mr. E. L.
Goldenberg, Mrs. E. L.
Goldschmidt, Mr. George
Gracie, Colonel Archibald
Graham, Mr.

Reproduction of the original First Class Passenger List in the possession of Jack Thayer